A Guide to Surface Engineering Terminology

A GUIDE TO SURFACE ENGINEERING TERMINOLOGY

General Editor: EUGENIUSZ TYRKIEL

Consulting Editor: PETER DEARNLEY

THE INSTITUTE OF MATERIALS

IN ASSOCIATION WITH THE IFHT

Book no. 575
First published in 1995 by
The Institute of Materials
1 Carlton House Terrace
London SW1Y 5DB

Technical publication produced
under the auspices of the
International Federation for Heat Treatment
and Surface Engineering (IFHT)

ISBN 0 901716 64 2

Typeset by The Institute of Materials
from disks supplied by P. A. Dearnley
Halftone origination by
Bourne Press, Bournemouth
Printed and bound in the UK by
Bourne Press

About the Editors

Professor Tyrkiel is in the Institute of Materials Science, Technical University of Warsaw, Poland and is former Chairman of the IFHT Terminology Committee.

Dr Dearnley is in the Department of Chemical & Materials Engineering, University of Auckland, New Zealand and is on the editorial panel of the journal *Surface Engineering* published by The Insitute of Materials.

Disclaimer

Editor's note

When referring to sub-atmospheric pressures, the torr (1 mm Hg) has been used throughout. It should be noted that 760 torr = 1 atm. = 1.03323 kg mm^{-2} = 14.6959 p.s.i Also 1 bar = 10^5 Pa = 0.986923 atm.

Introduction

The International Federation for the Heat Treatment of Materials was founded in 1971. The name was modified in 1986 to International Federation for Heat Treatment and Surface Engineering, but the acronym IFHT was retained.The objective of IFHT is to serve as the premier organisation for facilitating and promoting the international exchange of information in the science and technology of heat treatment and surface engineering. Its particular strength is its effectiveness as a forum uniting science and industry.

Surface Engineering is now accepted as a discipline in its own right. Although founded in humble origins, it presently enjoys a high level of technical sophistication. The technology which made this possible has largely come of age within the past 20 years. Surface Engineering has successfully transcended the 'patch-up technology' mind-set of the 1950s and 1960s, and is presently becoming increasingly linked with design and manufacture.

The IFHT has long been recognised as a leader in the promotion of internationally unified terminology in the field of heat treatment as reflected in the publication of a 'Multilingual Glossary of Heat Treatment Terminology' by the Institute of Metals (now The Institute of Materials) in 1986.

The IFHT in adopting the modern discipline of surface engineering has sought to provide an equally authorative 'Guide to Surface Engineering Terminology' based on the definition,

> 'Surface engineering involves the application of traditional and innovative surface technologies to engineering components and materials in order to produce a composite material with properties unattainable in either the base or surface material. Frequently, the various surface technologies are applied to existing designs of engineering components but, ideally, surface engineering involves the design of the component with a knowledge of the surface treatment to be employed'

The eminent multilingual materials scientist Eugeniusz Tyrkiel was invited to steer a Terminology Committee which initially solicited many terms from various industrial and academic experts from around the globe.Dr Peter Dearnley in his role as Consulting Editor has added further terms and shaped the material into a unified work. He is accountable for the presentation style and the illustrations therein. The work strives to be more than a mere collection of definitions. Wherever possible he has expanded the entries to provide greater technical detail and, where appropriate, cited further references and application examples. It is hoped that the book will serve as an informative quick-reference guide that will prove valuable to the expert and non-expert alike. I trust it will stimulate a wider appreciation and participation in the discipline of surface engineering.

Finally, I would wish to thank Professor Tyrkiel and his international team as well as Dr Dearnley and his family who sustained him through out his busy editorial task to provide an invaluable guide to surface engineering terminology.

Professor Tom Bell
University of Birmingham
March 1995

A

ablation. Vapourization or sublimation of a solid surface. Usually achieved by irradiating a solid surface with a high power density laser, plasma or electron beam.

ablative coating. (i) a special coating of limited durability applied to early types of re-entry space vehicles. The coating enables the space vehicle skin to remain relatively cool since most heat is consumed during heat of vapourization; (ii) a sacrificial coating applied to a metal to enable laser shock hardening. See *shock hardening*.

abradability. Proneness of a material to undergo abrasion.

abrasion. Material loss (wear) of a surface due to micro-scale cutting/shearing by impinging high hardness particles or grit. Also see *three body wear* and *erosive wear.*

abrasion resistance. See *abrasive wear resistance*.

abrasive blast cleaning. Also termed blast cleaning or abrasive blasting. A mechanical cleaning technique whereby a surface is exposed to the abrasive action of a stream of metallic or ceramic particles borne by high a velocity fluid (often air or water).

abrasive blasting. See *abrasive blast cleaning*.

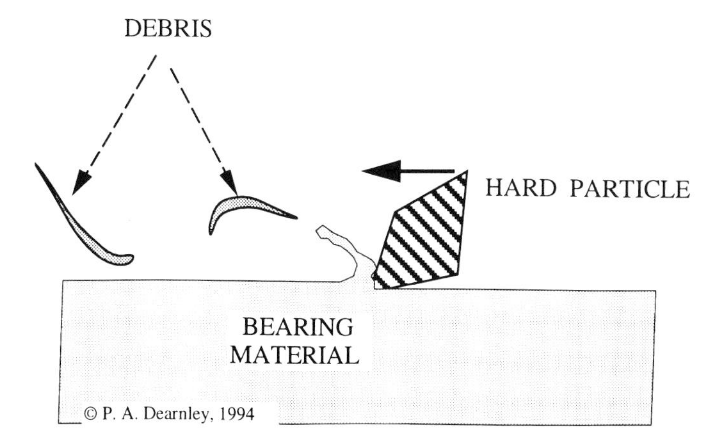

abrasive wear. Wear taking place by the mechanism of abrasion. A process which involves hard, fine particles, cutting through or ploughing over a relatively ductile bearing surface during sliding contact, thereby removing material from the surface (diagram). The abrasive particles maybe suspended in a fluid. Also see *three-body wear.*

abrasive wear resistance. Ability of a material to withstand abrasion.

acid cleaning. Chemical cleaning in which an acidic solution is applied.

acid descaling. Acid immersion process aimed specifically at the removal of oxide scale; also known as pickling in the steel industry where it is applied to hot rolled material prior to cold working. Also see: *alkaline cleaning, alkaline descaling* and *alkaline deruster salts.*

acrylic coat. Thermoplastic or thermosetting polymeric coating containing an acrylic polymer as the binder. They demonstrate good mechanical properties and high corrosion resistance. Thermoplastic variants are applied to cold rolled aluminium and steel sheet forms.

activated reactive evaporation (ARE). One type of plasma assisted physical vapour deposition (PVD); first popularised by Bunshah et al in the 1970's for the reactive deposition of ceramic coatings. The process is based on *reactive ion plating.* An anode is placed above an electron beam evaporation source enabling a high population of metal ions to be created (through electron-neutral atom/molecule collisions); the ions subsequently bombard the (usually) negatively biased substrates. This technique is one method of producing a *supported glow dischage.* However, it has failed to take hold as a significant industrial process; being superseded by the triode ion plating method.

activation analysis. See *NRA.*

activation overpotential. Part of the overpotential (polarisation) that exists across the electrical double layer at any electrode-electrolyte interface, which governs the rate of the electrode reaction by changing its activation energy.

activator. Commonly an alkali metal halide, an ammonium halide or alkali metal carbonate, which when added to a pack medium, serves to facilitate mass transfer from the source solid to the surface. For example, barium and calcium carbonates are added to hard wood charcoal to facilitate carbon transfer in pack carburising. NH_4Cl serves a similar purpose in pack aluminising and pack chromising. Another term for an activator is "energiser".

active medium. An essential component of a laser. See *laser.*

additon agent or additive. A material added to an electroplating solution to modify its operational characteristics or the properties of the resulting electrodeposit.

adherence/adhesion. A qualitative term representing the ability of a coat to form an intimate and durable attachment to a substrate. Chemical or metallurgical bonding across the coat/substrate interface offers the best adherence, although some improvement in adherence is offered by mechanical interlocking achieved by roughening the substrate surface (e.g., by grit blasting) prior to deposition. The latter is a common surface preparation prior to application of thermal spray coatings, like plasma sprayed ceramics.

adhesive (or adhesion) strength. Stress required to pull off or loosen a coating from its underlying substrate.

adhesive (or adhesion) strength tests. Several tests exist, but interpretation remains difficult. Such tests are not to be compared with established mechanical tests, like uniaxial tensile testing of bulk samples or indentation hardness. Available tests include scratch adhesion tests, indentation adhesion tests, peel and bend tests. Simpler tests involve applying a tape or glued rod end to the coated surface, which is subsequently pulled away causing partial or complete coating detachment; by knowing the force required to cause separation and the contact area under load, the stress to cause failure is calculated. Unfortunately, failure is not always at the coat-substrate interface; in such cases, the result does not equate to coating bond strength. Hence, caution is always necessary in interpreting results. Also see *scratch adhesion test.*

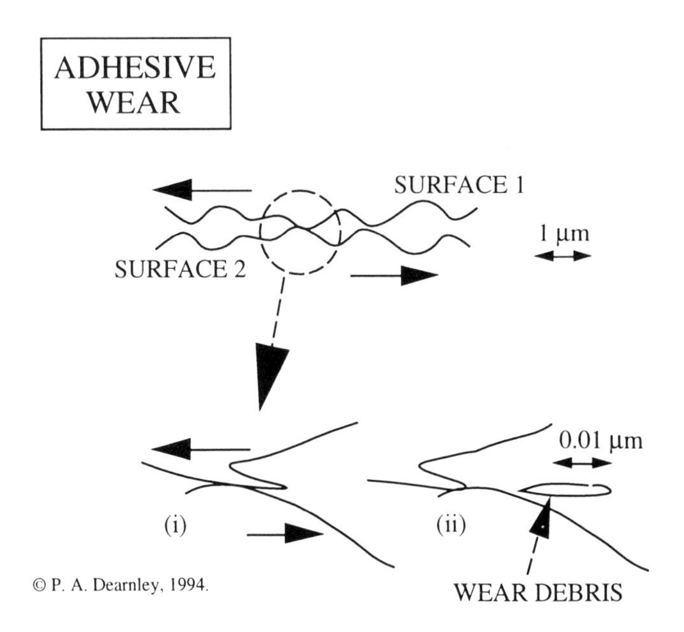

adhesive wear. A mechanism of wear involving localised welding together of micro-aperities during sliding contact, as for example between two bearing surfaces; such micro-seized contacts subsequently fail through micro-shearing or tearing causing the removal of surface material (wear). Debris particles can range in size from a few hundred micrometers to significantly less than one micrometer. Some tribologists now regard the term 'adhesive wear' to be redundant.A more common mode of wear debris creation (diagram) has recently been advanced by B. S. Hockenhull, E. M. Kopalinsky and P. L. B. Oxley (*J.Phys. D: Appl. Phys.*, 1992, **25**, A266-A272). These workers have produced a rigorous mechanical model of the interaction between rigid and ductile surface asperities during sliding contact . The model could be developed further to consider the more complex situation of the interaction between multiple asperities with similar plastic properties, as often arises in many engineering components.

adhesive wear resistance. Ability of any solid surface to resist adhesive wear.

adsorption. Bonding of atoms or molecules to a solid surface. Depending on the character of the bond involved, adsorption is subdivided into physical adsorption and chemisorption. The specie bonded to the surface is called an adsorbate, whereas the host solid surface is termed an adsorbent.

AES. See *auger electron spectroscopy.*

air plasma spraying (APS). See *plasma spraying.*

air spraying (of plastic coats). A method of spray forming plastic coatings; plastic particles are liquified (in-flight) by a thermal torch and propelled (by means of compressed air) onto the surfaces of components where they become rapidly solidified and consolidated to form a continuous surface coating.

A.L.E. See *atomic layer epitaxy.*

alkaline cleaning. Cleaning of metallic surfaces with an alkaline fluid (usually hydroxides, carbonates, silicates and phosphates, both complex and simple, of sodium) prior to electrodeposition.

alkaline deruster salts. For the removal of superficial rust on precision components without the danger of pitting or etching; especially suited to high tensile and hardened steels to achieve derusting without hydrogen embrittlement. The latter is possible if acid descaling treatments are used.

alkaline descaling. Alkaline cleaning process aimed specifically at the removal of scale.

alkyd coatings. Plastic coatings applied to metals to achieve stoved finishes. Alkyd coatings have high gloss and flexibility and, depending upon composition, can be of thermoplastic or thermosetting character. Hardness can be increased without compromising flexibility or adhesion by intermixing with urea or melamine formaldehyde while weather resistance can be improved by silicone resin additions. Thermosetting resins are used to achieve stove enamel finshes.

alloy plating. Any electroplating process in which two or more metals are simultaneously codeposited on the surface of an object forming a metallurgical alloy, e.g., see *brass plating* and *bronze plating.*

Almen number. A numerical value indicating peening intensity (during shot peening).

Almen test. Designed for measuring shot peening intensity. A standard sample, namely a flat piece of steel, is clamped to a solid block and exposed to the action of a stream of shot. The extent of curvature, after removing the sample from the block, serves as a basis for measuring the peening intensity. Peening intensity is influenced by the velocity, angle of impingement, hardness, size and mass of the shot pellets.

Alpha Plus. A proprietry gaseous austenitic nitrocarburising treatment. Also see *austenitic nitrocarburising.*

alternating current (AC) plasma CVD. A plasma assisted CVD (Chemical Vapour Deposition) in which a glow discharge plasma is generated by the application of an alternating bias potential. This represents a cheaper (although perhaps less satisfactory) alternative to the use of radio frequency (RF), microwave or pulsed plasmas. Such plasmas are essential when dielectric coatings (like oxides) are being deposited. They prevent electrical charge build-up on the component; this would happen if a DC (direct current) glow discharge plasma were deployed.

alumina coating. Aluminium oxide coating. Usually produced by CVD or plasma spraying. The former are fully dense and are frequently applied to cemented carbide cutting tools in combination with TiC and/or TiN and are up to 10 μm thick. These coatings usually comprise α-Al_2O_3 and/or κ-Al_2O_3. They often, but not always, exhibit marked crystallographic texture. Plasma sprayed alumina coatings are much thicker, generally ~100 μm, are microporous and contain α-Al_2O_3 and β– or γ-Al_2O_3. Such coatings have been used to provide wear protection for the edges of steel cutting blades used to produce wood-chips in the paper making industry; the chips are the raw material used to make pulp.

aluminising or calorising

'Diffusion metallising with aluminium' – IFHT DEFINITION.

Thermochemical diffusion treatment involving the enrichment of a metallic surface with aluminium in order to form layers enriched in metal-aluminide compounds. The process is mainly applied to nickel alloys and steels in order to increase their resistance to corrosion, oxidation and erosion. Aluminising can be subdivided into pack aluminising, *hot dip aluminising*, *salt bath aluminising* and *gaseous aluminising*. Pack aluminising has been widely applied for the oxidation protection of nickel base super alloys used in aero-engines; of crucial importance here is the diffusional formation of the intermetallic compound NiAl. Like many other pack cementation methods the pack comprises a source (Aluminium powder), an activator (NH_4Cl) and a diluent (Al_2O_3). One typical composition comprises (by weight) 15% Al (325 mesh), 3% NH_4Cl and 82% granular (120 mesh) Al_2O_3. It is essential to flood the pack with argon during pack aluminising, which for nickel base alloys is carried out ≈ 1090°C for 8 to 10 hours. Steels can be aluminised in the temperature range of 850-1050°C for 2 to 6 hours. If the mixture is tumbled shorter times often suffice. Diffusion layers maybe up to 100 μm thick. A secondary heat treatment can be carried out at 815 to 980°C for 12 to 48 hours in a neutral atmosphere. This improves the toughness of the aluminised layer and increases its depth (up to ~150 μm). The aluminising of austenitic stainless steels probably warrants further investigation. Apart from producing a hardened surface through the formation of metal-aluminides and aluminium oxide, aluminised austenitic stainless steels have the ability to act as hydrogen diffusion barriers. In this regard they show potential for use in fusion reactors. In the case of nickel alloys, pack aluminising technology is now being superseded to some extent by the increased use of NiCrAlY, CoCrAlY and other oxidation resistant coatings applied by thermal spray techniques, such as plasma spraying.

aluminium electroplating. A deposition process in which steel (usually) is coated with aluminium using an electrolyte comprising fused salts, e.g., 75% $AlCl_3$, 20%NaCl and 5% LiCl by weight operated at 170-180°C and 210 A/m^2. The process is relatively costly and is used only when other processes, applied for corrosion protection, are not practical.

Aluminium oxide coating. See *alumina coating.*

aluminium vapour plating. See *ion vapour deposition (ivadising).*

aluminosiliconising. The aim of the process is to increase heat (oxidation) resistance and, sometimes, corrosion resistance. Although mainly investigated for steels (see *minor thermochemical diffusion techniques*) there has been some recent interest in *laser alloying* Al + Si into the surfaces of titanium alloys. Such treatments have resulted in an improvement in the oxidational resistance of Ti-6Al-4V.

amorphisation (vitrification). The process of transforming a metallic or non-metallic material from the crystalline state into the amorphous or glassy state. Commonly in surface engineering, a surface is melted with a laser beam (using power densities ≈ 10^5 to 10^7 W/cm^2 with interaction times ~ 10^{-3} to 10^{-7} seconds) and rapidly solidified at cooling rates exceeding 10^5 K/s thereby suppressing the nucleation and crystallisation processes; this technique is termed *laser glazing.* Alternatively, materials (especially carbon and ceramic compounds) can be deposited as amorphous coatings (e.g.,via sputter deposition), if the substrate is kept at temperatures well below 100°C.

amorphous structure. An atomic structure lacking long range order or periodicity; characteristic of SiO_2 based glasses and other glassy solids. When irradiated by monochromatic X-rays such materials produce a characteristic broad peak spanning several degrees of 2θ, where θ is the Bragg angle.

Amsler wear test. A machine designed by the Swiss Company Amsler that enables two test wheels to run against each other with varying degrees of traction. Typically, the upper wheel is rotated at 90% of the peripheral speed of the lower wheel (diagram), although, many other configurations are possible. The test can be run lubricated or dry. The test was designed to emulate the contact stresses developed at the contact surfaces of gears. It has therefore proven useful in the optimisation of carburised and nitrided steels intended for use in gearing applications. Normal forces of up to 2000 N can be readily applied.

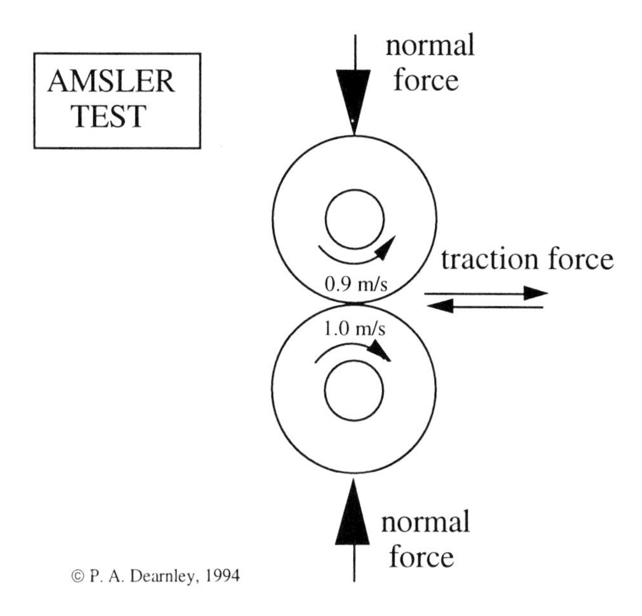

anion. A negatively charged ion that migrates towards the anode during electroplating or any DC plasma process.

anodic cleaning. See *electropolishing.*

anodic degreasing. A process of removing organic compounds, such as grease or oil, from the surface of an object by means of anodic polishing.

anodic oxidation. See *anodising.*

anodic polishing. See *electropolishing.*

anodic protection. A system of electrochemical protection whereby the corrosion rate of metallic components is reduced by making their potential sufficiently electropositive to render them passive. This is achieved by the application of an external voltage.

anodising. A treatment mainly applied to aluminium, magnesium and titanium alloys, whereby the natural passive surface oxide of these materials is thickened considerably by a process of anodic oxidation conducted in several proprietry electrolytic solutions. The oxide layers offer some durability against wear and corrosion and can be coloured by dyes for aesthetic appeal. One of the highest volume products is extruded aluminium forms used in window frames and other architectural applications. In recent years anodising has been exploited as a means of producing alumina based membranes for use in separation of gaseous mixtures in the chemical processing industries.

antifriction coating. Any coating that can be applied, especially to bearing surfaces, for the reduction of sliding friction and wear. Traditional metallic coatings used in lubricated bearing systems include lead, indium, silver and tin alloys. For dry lubrication systems, such as those used in satellite mechanisms (durability in outer space), ceramic coatings like TiC have found favour. Other low friction ceramic coatings, like TiN, ZrO_2 and Al_2O_3, are being evaluated for the bearing surfaces of artificial hip and knee joints.

antimonising

'Diffusion metallising with antimony' – IFHT DEFINITION.

Sometimes applied to bearings to improve running-in behaviour. See *minor thermochemical diffusion techniques.*

antiwear. A general term qualitatively denoting the ability of a material to resist wear. The term "antiwear" has become very popular within the 1990's as indicated by a recent conference series (held within the United Kingdom) that carry this name.

APS. See *plasma spraying.*

arc deposition. See *arc source PVD*

arc evaporation. See *arc source PVD*

Archard's wear equation. An early attempt (circa 1953) to produce a unified theory of wear. His equation is given as:

$$V = (KLS)/H$$

V = wear volume, K = wear coefficient (dimensionless), L = load, S = sliding distance; H = hardness; where all the parameters refer to the material being worn. Sometimes the quantity K/H is more useful. This is given the symbol k and is called the dimensional wear coefficient. The weakness of this theory is that it does not account for the rate controlling WEAR MECHANISMS and assumes that K remains constant regardless of the magnitude of L and S. Also see general comments made under *wear theory.*

arc source PVD. Also see *ion plating.* An evaporative PVD (physical vapour deposition) process in which (usually) the metallic constituent of the resulting ceramic coating is supplied through arc evaporation. A target plate of the required metallic component is repeatedly bombarded by high current density arc discharges, causing localised evaporation of very short (millisecond) duration (diagram). Because there is no stable molten pool, as is the case with electron beam source PVD, arc sources can be mounted in any orientation within a vacuum vessel. Modern designs use external magnetic fields to guide the arc in more efficient paths to enable unifrom consumption of the target plate (so-called steered arc technology). Other designs seek to minimise the generation of macro-particles; unwanted liquid droplets (~1-10 μm across) that are propelled at the component surface, solidify and become incorporated in the coating and act as sites of weakness when the coating is used in wear applications. Several approaches are used but they are all generally known as "filtered-arc" techniques. At the time of writing, this area continues to attract inventive ideas. See *filtered arc evaporator*.

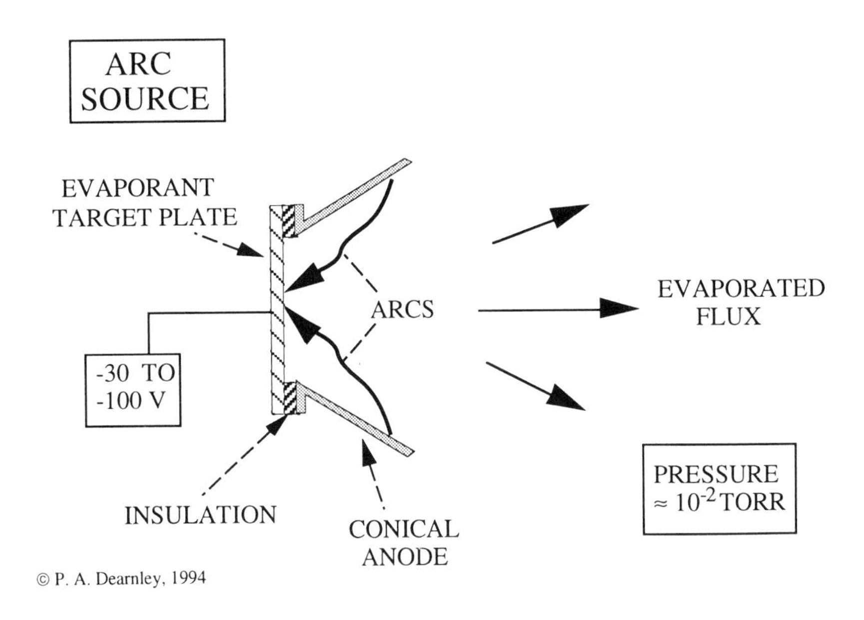

arc suppression. A feature incorporated into DC (direct current) power supplies used for high power density glow discharge devices, such as magnetron sputtering cathodes or plasma diffusion devices (e.g., plasma nitriders). The feature is able to detect the onset of an arc, indicated by a sharp rise in current and an equivalent sharp fall in voltage. On detecting these changes (which take place in the microsecond time range) the device momentarily shuts down the power supply and hence prevents the formation of detrimental current intensive arcs. Once an arc has been suppressed or "quenched" the device restores the operational electrical power, enabling the process to continue. With modern arc suppression technology, the total time from detection to full restoration of power can be as little as 60 μs, even for very large power supplies ≈ 500 kW. The diagram (p.18) is based on literature supplied by Advanced Energy Inc, U.S.A.

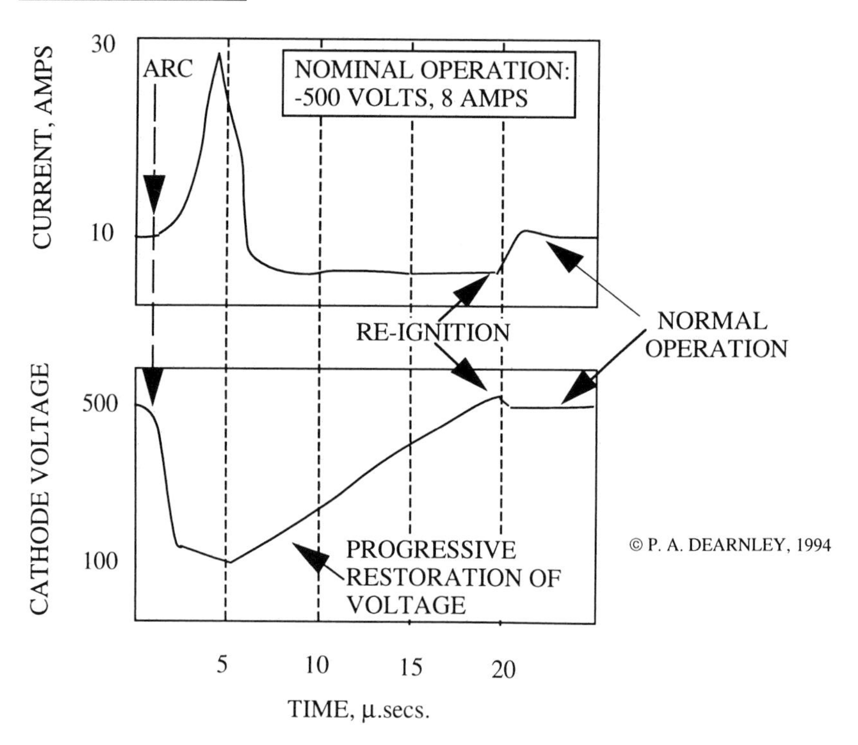

arc wire spraying. In this form of thermal spraying two wires of opposite electrical polarity are fed through a 'gun' on a converging path. Before they touch, a current intensive arc is created between them which liquifies metal from both wires. A blast of compressed air is passed across the molten pools of metal which causes them to 'atomise' into very fine droplets. Such devices have power ratings up to 15 kW enabling deposition rates of ~12 kg/h for stainless steel coatings. A diverse range of metallic coatings can be deposited, the most common being aluminium, copper and austenitic stainless steel. This torch is considerably noisier than *combustion wire gun spraying* but achieves higher deposition rates.

ARE. See *activated reactive evaporation.*

atmospheric CVD. See *normal pressure CVD.*

atomic layer epitaxy. Sometimes designated ALE. A comparatively new technique. The first experimental investigations were made in 1974 by Suntola and Antson, while the first industrial evaluation was made by the Lohja Corporation (Finland) in respect to the deposition of ZnS for use as electroluminescent displays. Other materials that are deposited include CdTe, GaAs, GaP, Ta_2O_5, and ZnSe. ALE seeks to deposit uniform, defect free layers of material that are only a few atomic layers in thickness onto oxide based substrates used for solid state electronic devices. The constituents of the coating are reacted precisely at the surface and become attached and formed through a repititive cycle of chemisorbtion and desorbtion; in the example of ZnS sequential atomic layers of Zn and S are alternately built up. For further details consult the review of Simpson et al, *Surface Engineering,* **3** (4), 1987, pp 343-348.

attrition wear. A term invoked by Trent to explain the wear of cemented carbide tools, whereby singular grains or groups of grains become torn or 'plucked' from the tool contact faces during metal cutting. Attrition wear is especially prevalent at relatively low cutting speeds when the metal chip forms a seized deposit around the tool cutting edge, termed a "built-up-edge", which periodically breaks away from the tool, and takes with it fragments of tool material. Hence, the chip-tool contact is of an intermittent nature. Even at higher cutting speeds, where the built-up-edge no longer forms, attrition wear can take place on the flank faces of cemented carbide tools, especially those containing high volume fractions of (W,Ti,Ta,Nb)C phase or having a relatively coarse WC grain size (≈3-5 μm). For further details see E. M. Trent: *Metal Cutting,* 3rd edn, Butterworths,London, 1991.

auger electron spectroscopy (AES). An electron spectroscopy method useful for obtaining chemical compositonal information of the outermost 10 atomic layers or so of a surface. The technique is carried out under ultra-high vacuum conditions (~10^{-9} torr), and uses a low power density electron beam to 'probe' the sample surface. During irradiation by the beam, characteristic auger electrons are emitted by the sample. Depth profiling can be achieved by the conjoint action of an ion beam gun, which directs argon ions at the sample surface and strips the sample surface of, for example, the natural external passive layer. The sample must be electrically conducting to assure satisfactory coupling by the electron beam. Hence, the method does not lend itself to the interrogation of ionic or covalently bonded ceramic surfaces (since they are non-conducting).

austenitic nitriding. Nitriding of any ferrous object whilst in the austenitic state. The most common austenitic nitriding is the plasma nitriding of austenitic stainless steels.

austenitic nitrocarburising. A thermochemical treatment whereby nitrogen and carbon are diffused into a plain carbon steel surface at temperatures such that the diffusion zone becomes stabilised into the austenitic state (by the nitrogen) for a significant duration of the treatment, whilst the core remains ferritic. Such treatments are mainly carried out below the Fe-C eutectoid temperature (~723°C) and above the Fe-N eutectoid temperature (~590°C).

The principle of austenitic nitrocarburising was initially demonstrated using a cynide salt bath method, but the process has come of age with gaseous nitrocarburising. Following the diffusion cycle a rapid quench assures the development of a martensitic and/or bainitic case which provides mechanical support for the external ε-$Fe_{2-3}N$ layer. Components treated in this way are able to withstand large point contact loads that would defeat conventionally nitrocarburised surfaces. Minimal distortion arises because the core structure always remains in the ferritic state. Also see *Nitrotec C.*

autocatalytic plating. A more exact description for electroless plating. Deposition of a metal (typically nickel) achieved by reducing metal cations (held in solution) by a controlled chemical reaction in the absence of an externally applied electric field. Once an initial metallic deposit is formed, it serves to catalyse the reduction reaction (autocatalysis). Also see *electroless nickel plating.*

B

babbitting. Any process that results in the mechanical or chemical attachment of a lead or lead-tin layer to a steel shell, for example as used in auto-engine journals. When applied by combustion wire gun spraying (using a lead-tin alloy rod in conjunction with an oxyacetylene flame) layers up to 25 mm thick can be achieved with minimal operator skill.

balanced magnetron. One type of source deployed in magnetron sputter deposition. The balanced magnetron is a standard magnetron sputter cathode in which the outer magnetic field is balanced to match that of the inner field. Until the mid-late 1980's all magnetron cathodes were of this design. The term has been invoked to differentiate this type of magnetron from the more recently developed unbalanced magnetron. Balanced magnetrons are still widely used for coating semi-conductor devices but are less popular for the deposition of tribological coatings, since the substrate current density is low and auxiliary heating is needed to assure adequate coating adherence. Also see *unbalanced magnetron.*

ball cratering test. A controlled polishing technique enabling the ready measurement of coating thickness; especially for thin (~ 1-10μm) ceramic coatings. A hardened steel or cemented carbide ball, ~ 20 to 50 mm diameter, is coated with 1μm diamond paste and rotated on the coated surface until the coating is worn through. Since the diameter of the ball is known, a taper section of known geometry is obtained. The width of the wear scar is easily measured with a light-optical microscope and the coating thickness calculated. The method gives good results for coatings that are 1 μm thick or greater.

BARE. Biased activated reactive evaporation - a PVD process; the usual form of activated reactive evaporation (see *activated reactive evaporation*).

barrelling. See *tumbling.*

barrel plating. An electroplating process devised for coating large numbers of small components (usually of relatively low mass), e.g., screws and fasteners; they are contained in a rotating cylinderical cathode, and in this way attain a coating of high uniformity.

barrier coating or layer. See *bond coating*

base material. See *substrate material.*

beam alloying. See *power beam surface alloying.*

bearing pressure. See *contact stresses*

bearing shells. Used for protection of the main and big end bearings of petrol and diesel engines. In the latter application very high bearing pressures must be sustained. Such bearings comprise a very ductile annealed low carbon steel substrate covered firstly by a layer (~100-200μm thick) of leaded bronze or copper-lead applied by casting or powder metallurgy means (see micrographs on p. 23). This layer itself is then electroplated with an overlay coating comprising various combinations of indium, lead, tin-lead and tin, totalling some 30μm thick. Indium or nickel serve as a bond coating between the overlay and the copper-lead bearing layer. See *lead plating, indium plating, tin-lead plating and tin plating.*

Berghaus, Bernhard. (See picture overleaf) The father of modern surface engineering who pioneered the exploitation of glow discharge plasmas for plasma nitriding and sputter deposition. His patents, originating during the 1930s and continuing through to the 1960s can be regarded as truly seminal. He was born of German parents in Amsterdam on 31st July 1896 and died in Zurich on 30th December 1966. Much of his early life was spent in Munster, N.R.W., Germany. He subsequently established a small research and development company called Gesellschaft zur Forderung der Glimmentladungsforschung whose sole goal was to achieve the industrial exploitation of plasma assisted techniques. Further impetus was provided by a susbsequent collaboration with Ionon of Köln (Cologne) which in turn lead to the foundation of Klockner Ionon who transferred plasma nitriding on a major scale to the European and North American manufacturing sectors. In later years many other companies emulated the ideas of Berghaus, and today plasma techniques, especially plasma nitriding, is practised in all technologically advanced countries. Klockner Ionon have recently merged to form Metaplas Ionon GmbH, Bergisch Gladbach. Information kindly provided by Dr Fritz Hombeck, former Managing Director of Klockner Ionon.

Berkovich indentation hardness. Hardness determined using a triangular base pyramid diamond indenter. The included angle at the apex of the pyramid is 65°, i.e., the angle between the vertical axis and each of the three faces. The Berkovich indenter is proving popular in *nanoindentation hardness.*

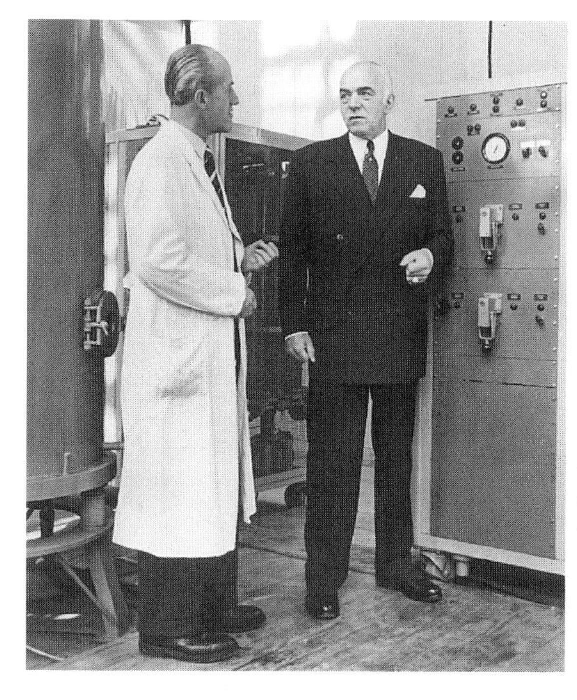

Bernhard Berghaus (1896–1966). *The father of modern surface engineering pictured (right) with one of his research staff. They are standing before a prototype industrial plasma nitriding chamber and controller, circa 1960. Courtesy of Metaplas Ionon GmbH, Bergisch Gladbach, Germany.*

berylliumising

'Diffusion metallising with beryllium' – IFHT DEFINITION.

The aim of the process, applied to ferrous as well as non-ferrous metals and alloys, is to increase the heat resistance of the object and to protect it against the corrosive action of molten metals. See *minor thermochemical diffusion techniques.*

Beta. A proprietry gaseous austenitic nitrocarburising treatment.

bias sputter deposition. The usual configuration of sputter deposition, whereby a negative bias is applied to the substrate during sputter deposition, the purpose of which is to raise the substrate ion density, causing heating and much improved coating adherence. The chamber walls are held at positive-ground potential. The magnitude of the substrate negative bias depends upon the type of sputter source and the point of time in the coating cycle. At the commencement of the treatment, a large negative bias (~1000 volts) maybe applied to effect a 'clean-up' of the components surfaces, through glow discharge sputtering. At the commencement of sputter deposition, the substrate bias voltage is reduced: (i) for a conventonal balanced magnetron source to ~500 volts; (ii) for a modern unbalanced source to ~ 50–200 volts. Also see *balanced magnetron, unbalanced magnetron* and *radio frequency (RF) bias sputter deposition.*

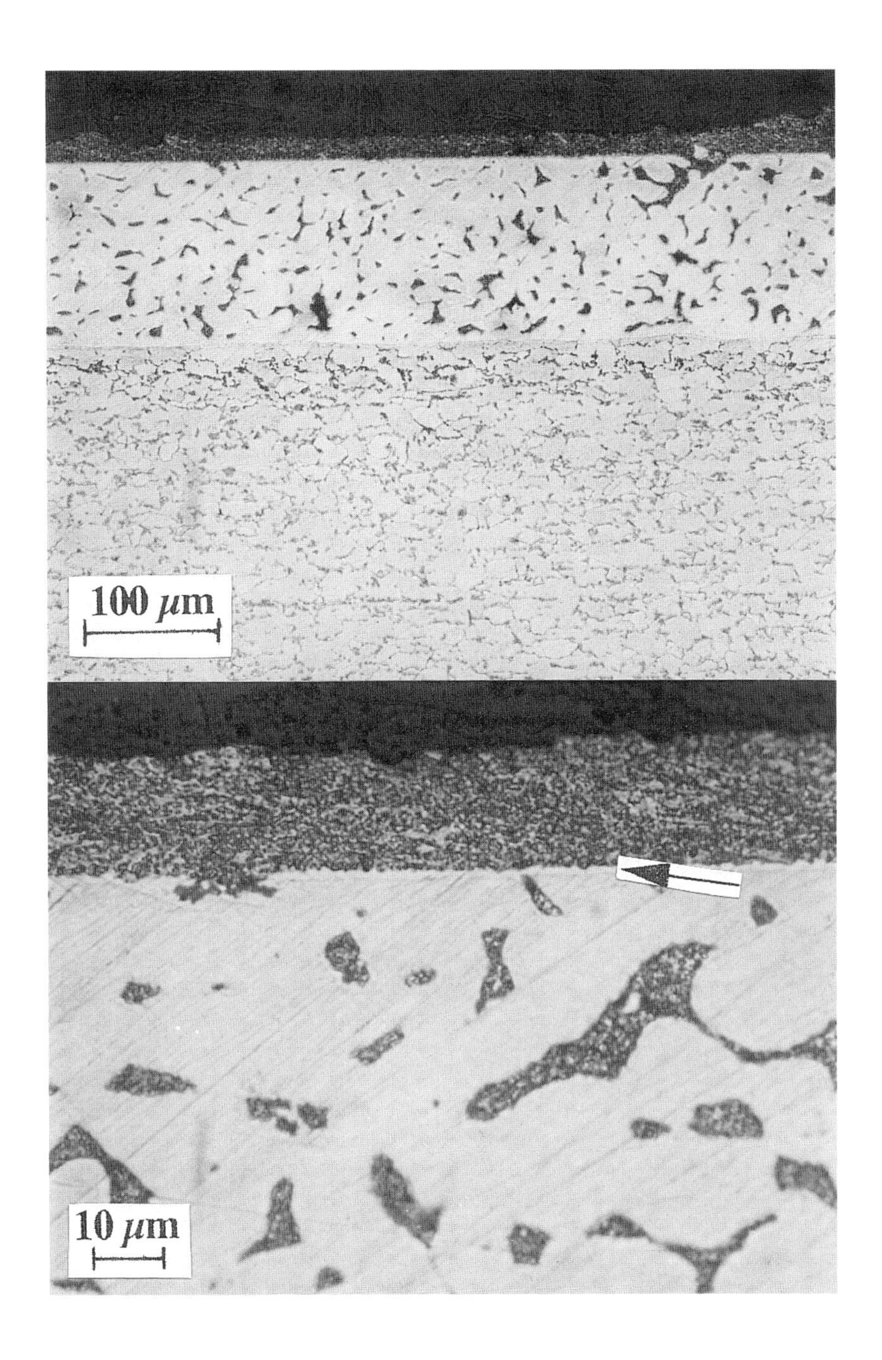

bearing shells. *Metallographic section of a big-end bearing shell designed for use in a high power diesel engine: (a) low magnification view showing total bearing layer on annealed low carbon steel shell; (b) high magnification view showing detail of outermost electroplated lead overlay coating and two-phased copper–lead alloy bearing layer. Note the indium bond layer (arrowed). Light optical micrographs (etched in 2% nital). From P. A. Dearnley, unpublished research, 1994.*

bimetallic corrosion. Also called galvanic corrosion. The corrosion produced when two dissimilar metals or alloys are in electrical (metallic) contact. One of the metals serves to stimulate corrosion in the other, which depends upon their respective corrosion potentials (*not* their standard electrode potentials); it is the metal which develops a more negative corrosion potential that becomes preferentially corroded. The current passed between the dissimilar metals is called the galvanic current, I_{galv}.

binder. (i) The main component of a polymeric coating material, which binds the pigment and filler particles together and creates a durable bond with the substrate. Also known as a film former. (ii) The metallic Co phase in WC-Co spray deposited coatings.

bipolar anode. A special variety of auxiliary anode in which current flow is not supported by an external connection.

blackening. A surface finishing process leading to the formation of a thin black oxide film on the surface of a metallic object, performed with the aim of improving corrosion resistance, usually for decorative applications. Steel objects are most frequently blackened in an oxidising bath or furnace. Also see *steam treatment.*

black nickel plating. An electroplating process in which nickel is deposited from baths containing zinc sulphate or zinc chloride additives; these produce dark, non-reflective decorative surface coatings.

black nitrocarburising. See *nitrotec.*

black oxide finishing. See *blackening.*

black oxide treatment. See *blackening.*

blank carburising

> 'Simulation treatment which consists of reproducing the thermal cycle of carburising without the carburising medium'
> – IFHT DEFINITION.

blank nitriding

> 'A simulation treatment which consists of reproducing the thermal cycle of nitriding without the nitriding medium'
> – IFHT DEFINITION.

blast cleaning. See *abrasive blast cleaning.*

blueing

> 'Treatment carried out in an oxidising medium at a temperature such that the polished surface of a ferrous object becomes covered with a

thin continuous adherent film of blue-coloured oxide'
– IFHT DEFINITION.

boiling white process. See *tin plating.*

bond coating or bonding layer. Any coating that serves to improve adherence. Commonly exploited in thermal spray coating systems. For example, when applying an Al_2O_3 or Cr_2O_3 ceramic coating by plasma spraying, to a steel component, it is usual practice to firstly apply a bond coating of Ni-Cr or Ni-Cr-B alloy to assist adherence. The magnitude of detrimental residual stresses that develop in the oxide coatings is minimised by this procedure. Bond coatings also serve as barrier layers to enable coatings to be applied to chemically incompatible substrates, e.g., amorphous carbon coatings can only be satisfactorily applied to steel substrates by applying an intermediate bond coating of silicon; this prevents diffusion of carbon into the steel substrate.

bond strength. See *adhesive strength.*

boost-diffuse cycle. A technique of non-eqilibrium carburising to enable rapid carbon transfer. See *vacuum carburising* and *plasma carburising.*

boriding or boronising

'Thermochemical treatment applied to an object with the aim of producing a surface layer of borides' – IFHT DEFINITION.

This thermochemical treatment involves diffusing interstitial boron into (mainly) steel or cemented carbide components at a temperature such that highly durable surface mono- or multi-layers of transition metal borides are formed through a reaction between the boron and one or more metals in the substrate. For steels, treatment temperatures are in the range of 750 to 1050°C; for cemented carbides temperatures are ≈1000°C. Other substrates, like titanium and nickel base alloys can be treated in principle but industrial boriding of these alloys is comparatively rare. Various boriding media are available: plasma, gas, salt bath (electroless and electrolytic), pack and paste. Salt bath media are mainly exploited in Eastern Europe; their negative environmental draw back has prevented wider exploitation in Western Europe, North America and Japan. The main-stay technique outside Eastern Europe remains the pack process, despite notable development efforts in gaseous and plasma boronising. A common pack comprises (by weight) 90% SiC, 5% B_4C and 5% KBF_4. The SiC acts as a diluent, the B_4C as the principal boron source while the KBF_4 serves as an activating and fluxing agent. The latter can be satisfactorily replaced by NH_4Cl. Boriding has the ability to convey improvements in wear resistance while simultaneously imparting resistance to corrosion by mineral acids and molten metals (especially zinc and aluminium). Boride layers produced through boriding tend to lack the toughness required for high contact loads especially those delivered at high strain rates; in this regard they are inferior to nitrided or carburised steels. The duplex FeB/Fe_2B produced on boriding steels, under moderate boron potential, results in the creation of detrimental residual elastic stresses in the vicinity of the

FeB/Fe_2B interface which can lead to cracking (see micrographs on p. 27). This is one factor that is responsible for the observed brittleness of borided surfaces. In the case of plain carbon and low alloy steels, boride layer toughness can be improved by boriding under low boron potential to produce a mono-phased layer of Fe_2B. Another possibility is to conduct a post boriding annealing treatment ≈850°C, in vacuo, thereby converting the duplex FeB/ Fe_2B into a mono-phased Fe_2B layer. The latter option is obviously less economic. For a recent review see H. J. Hunger and G. Trute, *Heat Treatment of Metals*, 1994, (2), 31-39. Also see *pack cementation.*

boroaluminising

'Thermochemical treatment involving the enrichment of the surface layer of an object with aluminium and boron' – IFHT DEFINITION.

See *multi-component boriding*

borochromising. See *multicomponent boriding.*

borocopperising. A comparatively rare multicomponent boriding diffusion method whereby a steel component surface is enriched in boron and copper. It is claimed that the resulting surface layer has all the positive attributes of an iron boride layer but with reduced brittleness. The treatment has been specified for steel components working under severe abrasive wear and, simultaneously, impact loading conditions. Pack borocopperising is the usual mass transfer medium. Packs comprise a mixture of B_4C and $CuCl_2$. The process is conducted at temperatures of 925-930°C for several hours. The thickness of the layer is in the range of 50 to 100μm.

boromolybdenising. Thermochemical multicomponent boriding treatment involving the silmultaneous diffusion of boron and molybdenum into the surface of a steel component. It is claimed that the resulting surface layer is more wear resistant than a conventional iron boride layer. The preferred mass transfer media used in Eastern Europe is by electrolytic and electroless salt baths. The molybdenum is added as Na_2MoO_4; Mo and B are transferred to the surface at the same time. The process is conducted at temperatures ~950-1100°C for 2-4 hours. It is not practised on a significant industrial scale.

boronising. See *boriding.*

borosiliconising. See *multicomponent boriding.*

borotitanising. See *multicomponent boriding.*

borotungstenising. As with boromolybdenising it is carried out on steel as an electroless or electrolytic salt bath diffusion method. The salts comprise a mixture of boron salts and Na_2WO_4. Boron and tungsten are simultaneously diffused into the steel. The microhardness of the resultant layer is ~ 2200-3000 kg/mm^2. It is rarely practised on an industrial scale.

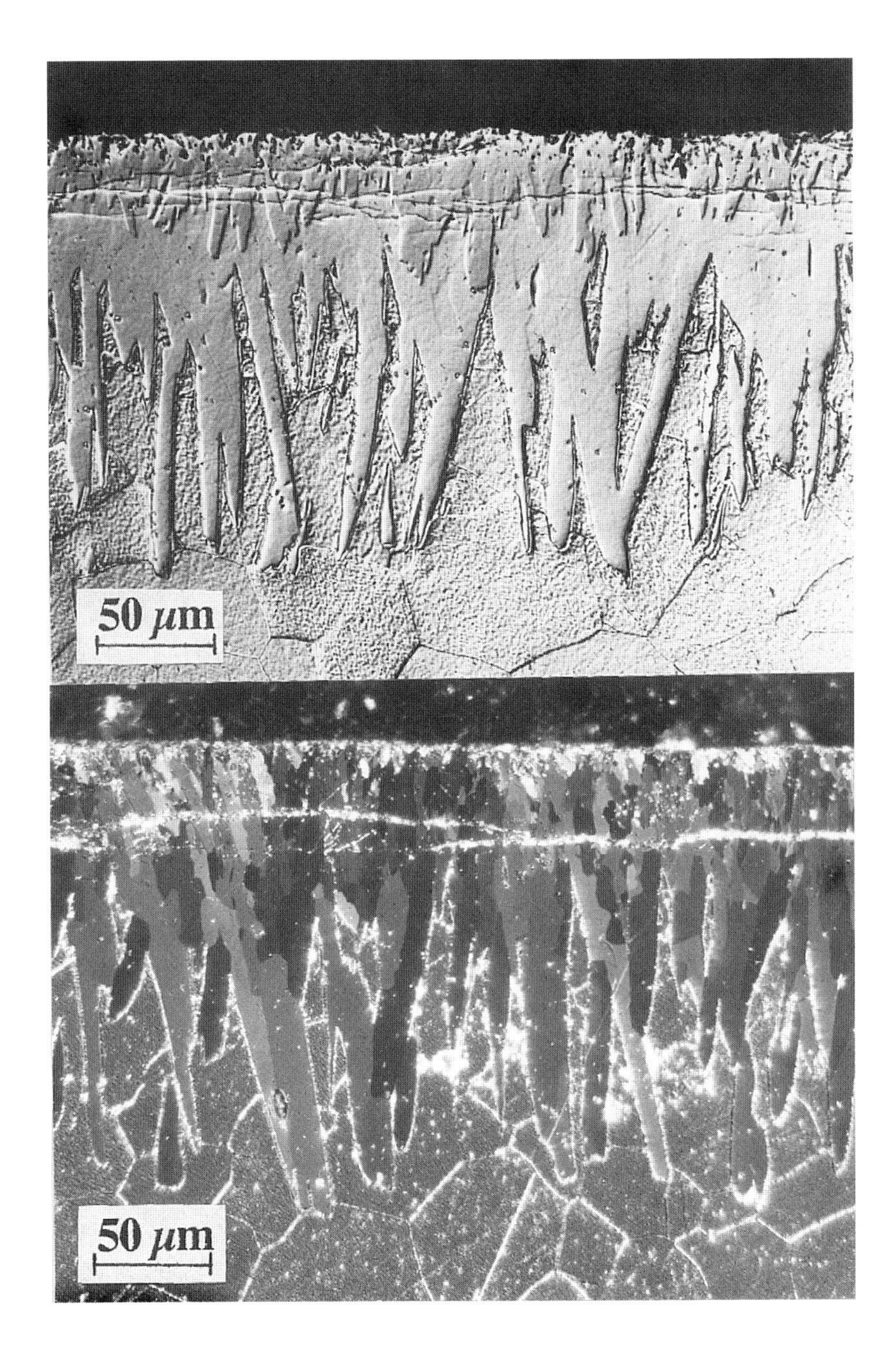

***boriding or boronising.** Normal metallographic sections of commercially pure iron following pack boriding at 900°C for 10h: (a) Nomarski interference contrast showing cracking in close proximity to the FeB–Fe_2B interface. The outermost layer is FeB, the adjacent layer Fe_2B; (b) the same sample viewed by crossed polarised light, revealing the coarse acicular grains of the boride layer (etched in 2% nital). Also see P. A. Dearnley and T. Bell,* Surface Engineering, *1985, **1**, (3), 203-217.*

Boudouard reaction. See *gaseous carburising*

boundary lubricant. Any lubricant that serves to enable stable boundary lubrication. If the lubricant fails, seizure will result. Boundary lubricants are often added to oils as "additives". Various types exist, which include hexadecanol, oleic acid amide and stearic acid.

boundary lubrication. A condition taking place in lubricated machine components where two sliding bearing surfaces pass over each other but are prevented from making metal on metal contact by a very thin boundary film of organic lubricant. It is argued that the film may only be a few molecules or even one molecule thick. This condition only arises under very high contact pressures or when sliding speeds are low. An example of this is the die-workpiece interface in wire drawing.

brass plating. An electroplating process in which copper and zinc are simultaneously deposited on the surface of an object. For decorative purposes a 60Cu:40Zn (by weight) alloy coating is preferred. It may be applied to steel, tin plated steel, zinc die-castings and even plastics. To prevent the brass plating from tarnishing it is usual to lacquer. Brass coatings also serve to enable good bonding between rubber and steel for applications such as shock absorbers and engine mountings. In this case a 70Cu:30Zn (by weight) alloy coating is used. The latter application is now, to some extent, being superseded by phosphating.

braunite. The eutectoid (2.35wt%N) decomposition product of Fe-N austenite, induced by appropriate cooling, after austenitic nitriding. It comprises α-Fe and γ'-Fe_4N

brightener. An additive made to electroplating or autocatalytic solutions which improves the specular reflectivity of the deposit.

bright finish. A surface finish of high specular reflectivity.

bright hardening

> 'Quench-hardening treatment involving heating of the object in a protective atmosphere so that the discolouration of its surface by oxidation is prevented' – IFHT DEFINITION.

brightness. Optical brightness at any point on a surface in a specified direction or orientation. Measured in candelas per unit area.

bright nickel plating. A nickel electroplating process in which nickel deposits containing from 0.02 to 0.13wt% sulphur are formed. Mainly applied as a primary coating before finishing with an outer chromium or, less commonly, precious metal electroplating.

bright nitriding

> 'Gas nitriding without the formation of a surface compound layer' – IFHT DEFINITION.

Plasma or gaseous nitriding of steels, carried out at sufficiently low partial pressure of nascent nitrogen to avoid the formation of a continous, exterior, "compound-layer" of γ'-Fe_4N and/or ε-$Fe_{2-3}N$, while providing sufficient nitrogen for the formation of a nitrided diffusion zone (or case) of engineering merit. Bright nitriding conditions are essential for critical components, like precision gears, where an external compound layer may flake-off to the detriment of service life. This effect has often been experienced with improperly gas nitrided components. If bright nitriding conditions cannot be achieved, compound layer can be removed by grit blasting or immersion in appropriate cyanide based solution. Also see *nitrogen potential* and *gaseous nitriding.*

bright plating. Electroplating under conditions that produce electrodeposits of high specular reflectivity.

brilliance. See *brightness.*

Brinell hardness. A ball of WC-Co is pressed by a heavy load (often 3,000 kg) into the surface of a metal and the diameter of the depression is measured. The Brinell Hardness Number (BHN) is the ratio of the load (kg) to surface area (mm^2) of the indentation, hence the units are expressed in kg/mm^2. Both macro and microindentation versions exist. The technique is widely used in the steel heat treatment sector. Named after J. A. Brinell (1849-1925).

bronze plating. An electroplating process in which copper and tin are simultaneously deposited on the surface of an object to form a bronze coating. Alloy coatings containing 10-20 wt-% tin are typical. The anode should be oxygen free high conductivity copper, often in the form of 'slugs', contained in a titanium anode cage and operated at a current density not exceeding 2 amp/dm^2. Tin is provided from the electrolyte, which is added as potassium stannate. The copper content of the electrolyte is maintained by anodic dissolution. Bronze plating has an attractive golden appearance that tarnishes unless lacquered. The technique is mainly applied for decorative purposes. Also see *speculum plating.*

brush plating. A method of electroplating in which the surface of the object to be plated is made cathodic. The object is wiped by a pad or brush connected electrically to the anode and moistened by an electrolyte. The technique is used for the restoration of worn or damaged components which are often too complicated or too massive to be factory electroplated.

buffing. A surface finishing operation performed for producing a smooth, reflective, surface by bringing it into contact with a revolving soft-fibre wheel charged with a suitable compound consisting, typically, of a fine abrasive immersed in a binder.

burnishing. A mechanical surface finishing operation applied to metallic objects. The surface is subjected to a micro-hammering action by one or more hardened polished steel tools making a rotary or reciprocating motion. The main purpose of burnishing is to rapidly obtain a very smooth surface. It work-hardens the surface and increases the fatigue strength of the component. Burnishing may inadvertently happen during surface grinding if the selected operational parameters are too severe for the material being ground.

burnt deposit. A rough, poorly adherent electrodeposit formed by the use of excessive current density.

burnt surface. A discolouration (oxidation) of the surface caused by grinding with a too severe metal removal rate.

C

cadmium plating. An electroplating process in which cadmium is deposited onto steels to improve corrosion resistance. Once widely used for aerospace applications, because of its superior fatigue properties (compared with standard electroplatings) it is nowadays in a state of comparative decline because of the high toxicity of cadmium. This factor stimulated the growth of alternative treatments such as aluminium coating by the *Ivadising* process.

calcarous scale. A calcium carbonate-magnesium hydroxide enriched deposit frequently deposited from hard water.

calorising. See *aluminising.*

carbon activity. When carburising steel, it is the ratio of the vapour pressure of carbon in austenite to the vapour pressure of graphite (the reference state) for any given temperature.

carbon availability

> 'Amount of carbon in grams per cubic metre of gas which at a given temperature can be transferred to the surface of an object, while the carbon potential decreases from 1 to 0.9 weight %'
> – IFHT DEFINITION.

carbon dioxide (CO_2) laser. A laser in which the active medium comprises a mixture of 10% CO_2, 30% N_2 and 60% He. CO_2 laser light has a wavelength of 10.6 μm. For laser alloying and transformation hardening, carbon dioxide lasers are typically rated in the range of 1 to 3 kW and can deliver a maximum power denisty $\approx 10^6$ to 10^9 W/cm^2 to the surface. For further information see *laser.*

carbonitrided case

> 'Surface layer of an object within which the carbon and nitrogen content has been increased by the carbonitriding process – IFHT DEFINITION.
>
> Also see *case depth.*

carbonitriding. An austenitic thermochemical treatment applied to steels in which carbon and nitrogen are simultaneously diffused into the surface; carried out within the lower temperature range of carburising, usually not above 900°C. Higher treatment temperatures result in a marked reduction in nitrogen up-take (see comments on *salt bath carbursing*). Following the diffusion cycle, components are oil or gas quenched to develop a martensitic case. Since treatment temperatures and times are generally lower than for carburising, case depths are shallower (up to 0.75 mm). Carbonitriding is conducted using gaseous, or salt-bath methods. Plasma and vacuum carbonitriding have not been industrially developed. In the case of salt-bath treatments, sometimes called cyaniding, there is essentially no distinction from salt bath carburising, except in respect to treatment temperature and time, as indicated earlier. Carbonitriding is less energy intensive than carburising which makes it economically more attractive. Nitrogen has two effects: (i) hardenability of the case is increased; (ii) the M_s temperature is depressed. Hence, it is possible to achieve satisfactory case hardnesses when treating non-carburising grades of steel, e.g., plain carbon steels. Such steels are problematic for conventional carburising, where there is no accompanying enhancement in hardenability. Carbon potential is of critical importance. For a given carbon content, carbonitrided cases exhibit a larger quantity of retained austenite than for equivalent carburised cases. It is also argued that the nitrogen serves to improve wear and temper resistance, over and above that which is possible through simple carburising.

NOTE: Carbonitriding should never be confused with nitrocarburising.

carbon potential

> 'The carbon content of a specimen of pure iron in equilibrium with the carburising medium under the conditions specified' – IFHT DEFINITION.

The amount of nascent carbon available at the surface for solution in austenite during carburising. Often expressed in wt-% in relation to the Fe–C system. Depending on the media employed, the carbon may or may not be in equilibrium with the surface. Measures of carbon potential can be gained from dew point determination, infra-red gas analysis or oxygen sensing. Also see *gas carburising, infra-red gas analyser* and *zirconia oxygen sensor.*

carbon profile. Carbon concentration as a function of depth below the surface.

carbon restoration.

'Carburising to replace surface carbon lost during prior heat treatment' – IFHT DEFINITION.

carburisation. See *carburising.*

carburised case.

'Surface layer of an object, within which the carbon content has been increased by the carburising process' – IFHT DEFINITION.
Also see *case depth.*

carburiser. Jargon; referring to a gaseous or plasma carburising vessel.

carburising.

'Thermochemical treatment involving the enrichment of the surface layer of an object with carbon' – IFHT DEFINITION.

Thermochemical diffusion treatment involving the enrichment of a surface with carbon. Mostly applied to low carbon (0.12-0.18 wt-%) steels, but non-ferrous metals, like titanium can also be hardened through carburising, but the mechanism of hardening in that example is quite different. Nowadays, carburising is mostly carried out using gaseous, fluidised bed or plasma media, although in some technologically less developed countries, salt bath, paste and pack methods are still widely used. This comment is reserved in part for large volume industrial operations and excludes small workshops and other engineering businesses where pack carburising is frequently justifiable given their very small turnover of work; in such cases, investment in major capital equipment is clearly unjustified.

Steels are generally carburised between 850 and 1050°C, whilst in the austenitic state; this stage is followed by an oil or gas quench (to ambient temperature) which causes the formation of martensite. Subsequently, the carburised steels are tempered at approximately 150 to 200°C, to obviate case embrittlement. The carbon content of the steel core is sufficiently low (0.12-0.18wt-%C) to retain relatively high toughness, even in the as-quenched state. Specified case depths vary depending upon the application loading, but are generally deeper than those attained by nitriding, reaching a maximum of approximately 2mm. Carburising, like nitriding, improves rolling contact fatigue endurance since the volume expansion accompanying the hardening step, places the carburised case in a state of residual compressive stress. The majority of drive shafts and gears used in automobiles are carburised. It is probably the singular most important surface engineering technology used in the automotive vehicle sector. See *gaseous carburising, pack carburising, plasma carburising, salt bath carburising* and *vacuum carburising.* With regard to process control, it is very important to achieve the correct carbon content in the carburised case (≈0.8 wt-%) otherwise excessive retained austenite will result. See *retained austenite.*

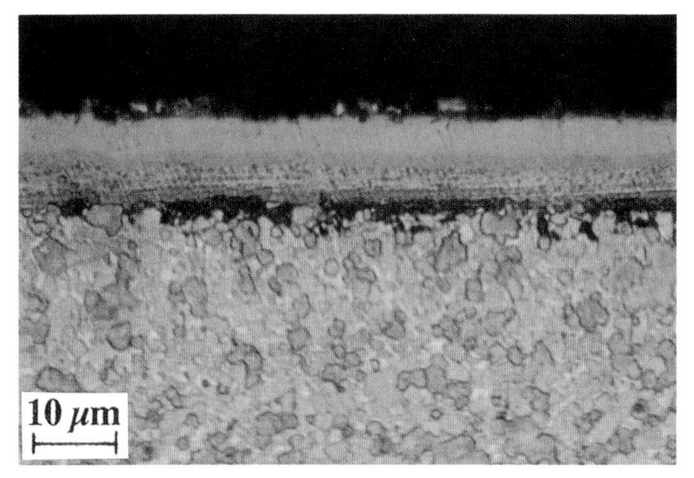

chemical vapour deposition (CVD). *A typical duplex layered CVD coating of TiN (yellow-orange) and adjacent layer of TiC (grey) on a steel cutting grade cemented carbide. When used to cut steels, the rake face wear resistance of TiN is slightly superior to TiC, whereas the flank wear resistance of TiC is consistently superior to TiN. A small amount of η-phase is also visible at the coat–substrate interface; nowadays, this is comparatively rare. Light optical micrograph; specimen etched in 10% alkaline potassium ferricyanide. From P. A. Dearnley, unpublished research, 1993.*

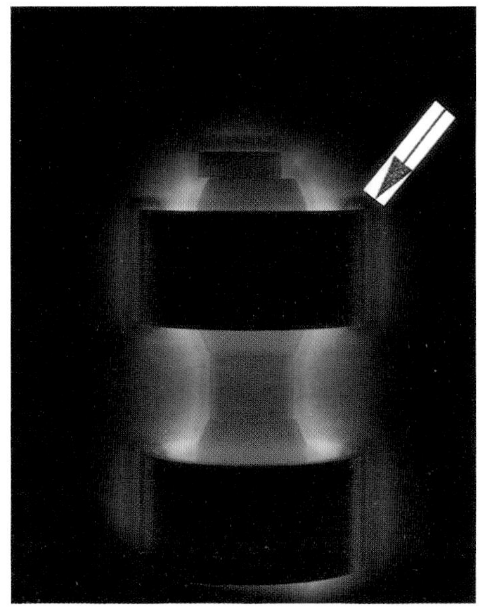

glow discharge plasma. *A small titanium component undergoing plasma nitriding. The anatomy of the glow discharge is partly revealed, showing the dark space or glow seam (arrowed). Notice that at this pressure (≈5 torr) the dark space faithfully follows the geometry of the component. From P. A. Dearnley, unpublished research, 1988.*

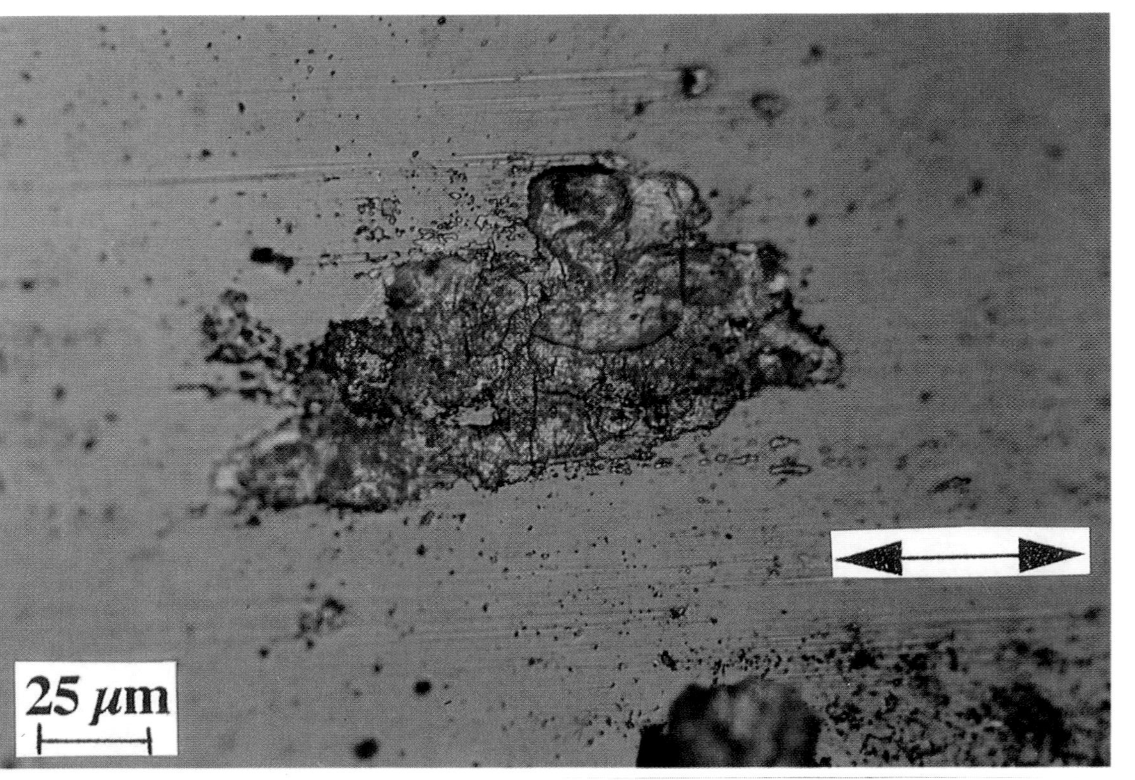

oxidational wear. *Light optical micrograph of the worn surface of a PVD–TiN coated titanium alloy test pin showing the formation of a localised oxide film generated by 'hot-spots'. The oxide exhibits interference colour fringes and is cracked. Note the characteristic smooth appearance of the background TiN (yellow); the aftermath of oxidational wear. The test pin has been reciprocated at 1 Hz against a TiN coated titanium disc (that showed similar wear features) under a normal load of 30N in a saline solution. The double arrow shows the direction of reciprocation. Light optical micrograph (unetched). From P. A. Dearnley, unpublished research, 1994.*

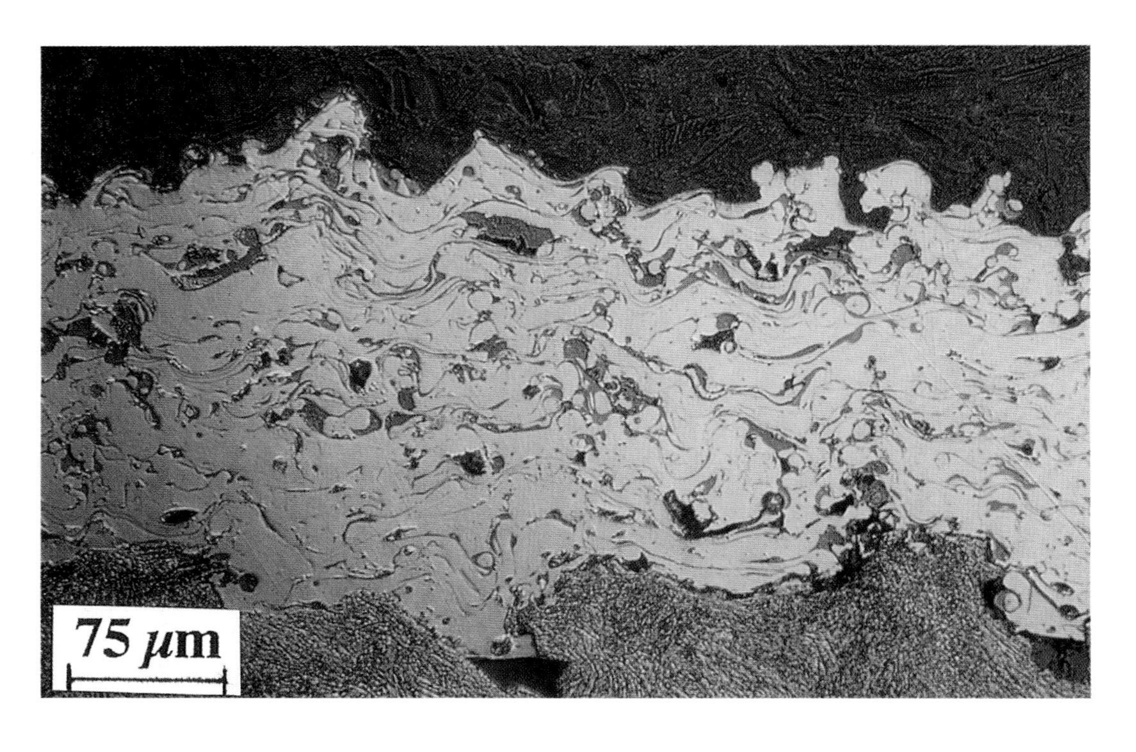

plasma spraying. *Metallographic section through a low alloy steel plasma sprayed with 316L in the open atmosphere. The dark inclusions in the coatings are oxides formed through oxidation of the molten 316L particles in-flight. The dark inclusions at the coat–surface interface are silicon carbide grit particles embedded in the steel substrate during grit blasting prior to plasma spraying. Note that the original steel surface (now the coat–substrate interface) was roughened by the grit blasting procedure, enabling keying-in of the coating. Light optical micrograph, Nomarski interference contrast, (etched in 2% nital). From K. L. Dahm and P. A. Dearnley, unpublished research, 1994.*

(a)

plasma nitriding. *Plasma nitriding is nowadays practised on a wide industrial scale: (a) a 1.5 m diameter industrial plasma nitriding chamber with computer control; (b) a steel food-stuff pelletiser undergoing plasma nitriding. Note the uniform coverage of the component by the glow discharge plasma. Photos courtesy of Dr F. Hombeck former Managing Director of Klockner Ionon, Leverkusen, Germany.*

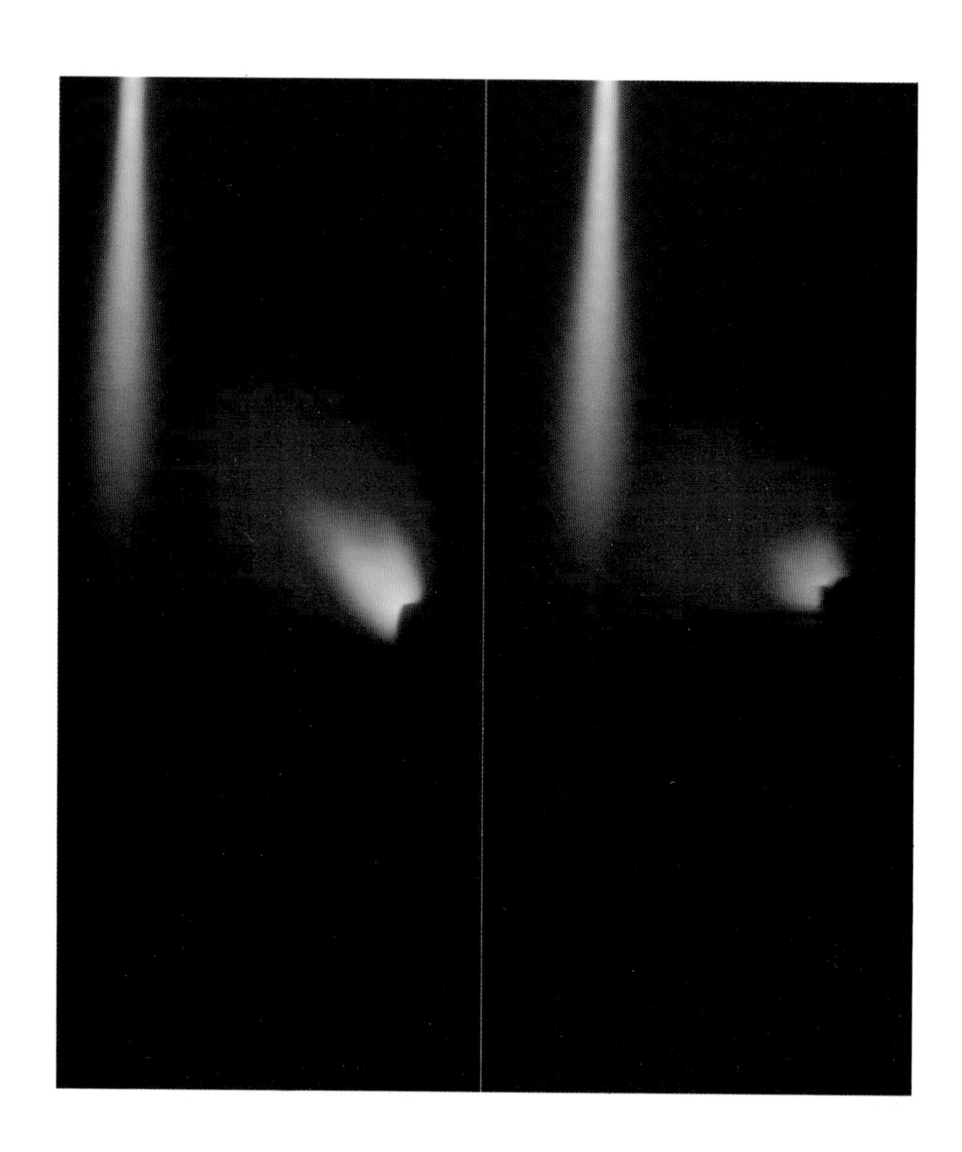

F

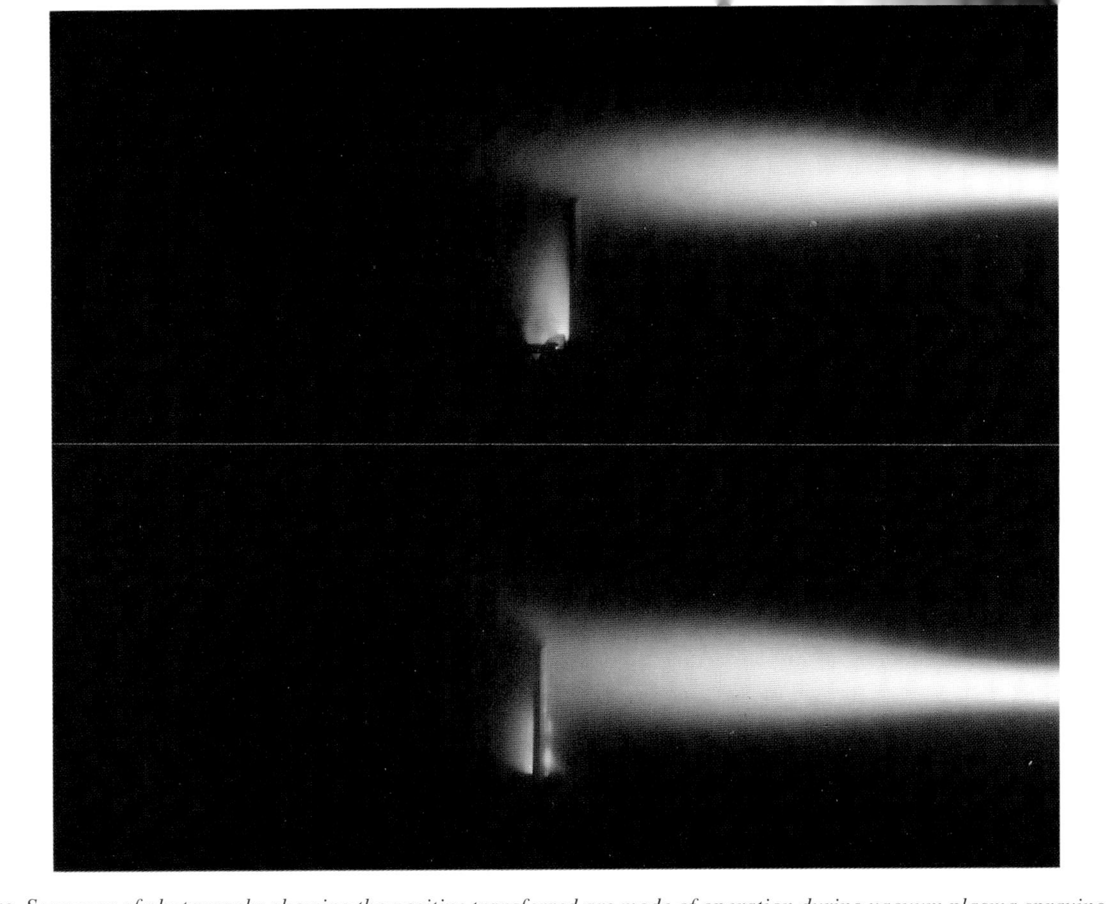

transferred arc. *Sequence of photographs showing the positive transferred arc mode of operation during vacuum plasma spraying (VPS). As the plasma plume (yellow-orange) approaches the substrate, a localised arc discharge is created (red), causing the substrate to be heated to temperatures above 300° C. During this stage, no deposition is allowed to take place. The procedure serves to 'clean up' the substrate, assuring a greater bonding of the subsequently plasma sprayed coating. From P. A. Dearnley and K. A. Roberts, Powder Metallurgy, 1991,* ***34****, (1), 23-32.*

vacuum plasma spraying (VPS). *Modern plasma spraying equipment, although highly automated, requires a high level of operator skill. The photograph shows Mr Kevin Roberts, one of the technical staff of Cambridge University, setting the carrier gas and powder feed rates (right hand control panel) of a modern VPS system, prior to commencing a coating run. The central control panel operates the CNC control system that achieves precise plasma torch manipulation, essential for maximising coating coverage of components. Photo by P. A. Dearnley, 1989.*

unbalanced magnetron. *Photograph of an unbalanced magnetron operating at the unusually high pressure of 20 mTorr of argon. The magnetron comprised permanent balanced magnets with an additional outer electromagnet providing the unbalance; the coil current was 5 A. Note the extension of the plasma out of the main plasma 'doughnut'. The target was a 120 mm disc of titanium. The colour of the discharge (turquoise) is a characteristic of sputtered titanium, caused by the excitation of in-flight sputtered titanium atoms. Despite using a very fast film (ASA 1600), large depth of field (f32) and high shutter speed (1/2000 s), the doughnut still appears blurred; compare the sharp focus of the magnetron housing pictured in the background. From P. A. Dearnley, unpublished research, 1994.*

carburising medium.

'Any medium capable of carburising an object under a given set of conditions' – IFHT DEFINITION.

Also see *gaseous carburising, pack carburising, paste carburising* and *plasma carburising.*

carburising steels. Low carbon steels (~0.12–0.18 wt-%) with sufficient nickel, chromium, molybdenum and manganese to enable a martensitic case to be produced by oil quenching *after* carburising in the usual austenitic temperature range. Among the most popular types of nickel-chromium-molybdenum steels are BS970:832H13, BS970:835M15 and AISI:8620.

carrier gas.

'Base gas of a reactive atmosphere' – IFHT DEFINITION.

case. The surface zone that has received a change in alloying content after a given thermochemical diffusion method, such as nitriding or carburising.

case depth.

'Reference parameter for determining the distance from the surface of a heat treated object to the case/core interface, as specified by a characteristic of the material' – IFHT DEFINITION

Strictly speaking this is the depth of diffusional penetration by, for example, carbon after carburising or nitrogen after nitriding. However, within the heat treatment sector, a number of definitions of case depth have emerged; some have become formalised by appropriate standards authorities, like ASTM, DIN and SIS. In the USA, case depth is frequently defined as that portion of the treated zone which has a hardness above 50 Rockwell C. The most precise measurements employ the microhardness technique with Vickers or Knoop diamond indenters. In this instance, case depth is frequently defined as the depth of case with a hardness that is at least 50 kg/mm^2 above that of the core.

case hardening. A general term meaning *carburising, carbonitriding* or *induction hardening.*

cathodic cleaning. Electrochemical cleaning in which the object to be cleaned constitutes the cathode in an electrochemical cell. Separation of contaminants from the surface is achieved by the scrubbing action of liberated hydrogen.

cathodic electrocleaning. See *cathodic cleaning.*

cathodic protection. A method of reducing corrosion rate by making the potential of the metallic object(s) requiring protection more negative with respect to, for example, sacrificial plates or blocks placed in strategic locations. The steel hulls of ships can be protected in marine environments by attaching sacrificial plates of magnesium (a relatively more anodic metal) which renders the hull cathodic and therefore passive. Another method uses external anodes (made from graphite, mixed metal oxide coated titanium or high silicon-chromium cast iron) connected to the positive terminal of an external DC (direct current) power supply, while the structure requiring protection is connected to the negative terminal. Current is made to pass from the anode to the cathodic structure through the surrounding natural electrolyte, such as ground water. The latter approach is termed impressed current cathodic protection.

catholyte. Portion of an electrolyte adjacent to the cathode

cation. A positively charged ion that migrates towards the cathode during electroplating or any DC plasma process.

caustic embrittlement. Stress corrosion cracking of plain carbon steels by the action of caustic alkaline solutions.

cavitation damage. The erosion of a surface caused by the collapse of vacuum bubbles formed in a fluid. A condition which frequently affects ship propellers and impellers.

cementation. (i) An obsolete term denoting carburising; (ii) A general term used to denote any surface diffusion method that results in the formation of a surface layer containing interstitial or intermetallic ccompounds. See *diffusion coating* and *diffusion metallising*.

ceramic coating. Any ceramic coating produced by thermal spraying, CVD, PVD or plasma assisted PVD.

ceramic diffusion coating. Any ceramic coating produced by a thermochemical diffusion treatment.

ceramic dip coating. A ceramic coating process in which the objects are dipped into a ceramic slip and subsequently subjected to drying and high temperature sintering/firing.

ceramic flow coating. A modification of ceramic dip coating, in which a ceramic slip is supplied onto converyorised objects. The slip flows from nozzles designed to flush all surfaces; excess fluid is drained into a catch-basin and recirculated.

chemical cleaning. Any cleaning performed by the chemical action of a fluid. The surface of the object is sprayed by, or immersed in, a fluid which may be an acid, alkaline or organic solvent.

chemical etching. See *etching*.

chemical polishing. Polishing of a metallic surface by immersion in a bath containing oxidising substances which levels and brightens the surface by preferentially dissolving projecting surface irregularities.

chemical vapour deposition (CVD). See ***colour section,*** p. A. Less commonly termed chemical vapour plating, thermochemical plating or gas plating. A reactive gas phase deposition process in which one of the reactants (typically a metal halide) is in the vapour state, prior to admittance into the reaction chamber. (Note: a vapour differs from a gas in that it can be condensed by the application of an external pressure. Conversely a gas, being above its critical vapour pressure, cannot be condensed in the same way.) CVD can be carried out at atmospheric or sub-atmospheric pressures. For the production of carbide and nitride ceramics, reactions are of the general form:

$$MCl_x + H_2 + 0.5N_2 \longrightarrow MN + xHCl$$

$$MCl_x + CH_4 \longrightarrow MC + xHCl$$

The kinetics of reduction of metal halides like $TiCl_4$ mean that the CVD of TiC, TiN, Ti(C,N) and Al_2O_3 requires temperatures ~1000°C at atmospheric (760 torr) or sub-atmospheric pressure (~50 torr). However, by ionising the reactive gases in a radio frequency (RF) glow discharge, the kinetics of reaction can be accelerated to enable deposition at substrate temperatures ~500-600°C. Also see *plasma assisted CVD.*

chemical vapour infiltration (CVI). A CVD process applied to porous or fibrous objects.

chemisorption. Adsorption in which the atoms of adsorbate and adsorbent are held together by chemical bonds, e.g., ionic or covalent bonds. Also see *physical adsorption.*

chipping. Micro-fracturing or breaking away of fragments of a brittle coat especially at an edge or a corner.

chromaluminising.

> 'Thermochemical treatment involving the enrichment of the surface layer of an object with chromium and aluminium'
> – IFHT DEFINITION.

Thermochemical diffusion treatment, carried out at 800–1100°C, involving the simultaneous or consecutive enrichment of metallic surfaces with chromium and aluminium. The aim of the process is to increase the heat (oxidation) and erosion resistance of ferrous, nickel and titanium alloys. Chromaluminising can be conducted using pack, paste, salt bath or gaseous media, of which the pack method has received most attention. See *pack chromaluminising.*

chromaluminosiliconising. The aim of the process is to increase heat (oxidation) resistance and erosion resistance of steels, superalloys and high-melting point metals and alloys. See *minor thermochemical diffusion techniques.*

chromating. Also called chromate conversion coating. This treatment can be applied to a diverse range of metals with the purpose of improving their resistance to atmospheric corrosion. It involves immersing objects in an aqueous solution of chromic acid or chromium salts such as sodium chromate or dichromate which react with the metallic object forming a protective surface chromate film. Specific chromating solutions have been formulated for specific metals or alloys. In the case of metals with inherently high corrosion resistance, like cadmium, zinc and some aluminium alloys, the chromate treatment is itself sufficient to provide improved corrosion resistance without further refinement. During in-service atmospheric exposure the chromate deposit becomes partially dissolved by any surface moisture; the corrosive action thereby being inhibited. Chromating of base metals like magnesium and iron is insufficient in itself to enable a notable improvement in corrosion resistance; it is however beneficial when used for example after phosphating and before painting, i.e., it makes a useful contribution to the total surface protection.

chromesiliconising. See *minor thermochemical diffusion techniques.*

chromising

'Diffusion metallising with chromium' – IFHT DEFINITION.

Thermochemical diffusion treatment involving the enrichment of plain carbon steel surfaces with chromium, to impart corrosion and wear resistance. Chromising is generally carried out between 850 and 1050°C for durations up to 12 hours, which produces chromised layers ranging in thickness from ≈10 to 150μm. The constitution of the chromised surface layer depends upon the carbon content of the steel. For steels with 0.3 wt-% C, the chromised layer is essentially ferritic (the chromium remains in solution). Steels containing 0.4 wt-% C show evidence of intergranular carbide precipitation, while steels with higher carbon contents exhibit layers enriched in massive carbide deposits of the same type, i.e., $(Fe,Cr)_7C_3$ and $(Fe,Cr)_{23}C_6$. The Vickers microhardness of chromised steel surfaces increase linearly from ≈600 to 1800 kg/mm^2 for steels with carbon contents ranging from 0.2 to 0.8 wt-%. The higher the carbon content of a steel, the thinner the chromised layer thickness (for a given chromising temperature and time). Chromising can be conducted using various media. See *pack chromising* and *gaseous chromising.*

chromium plating. An electroplating process in which chromium is deposited from a chromic acid solution in the presence of silicofluoride and/or sulphate catalytic anions. Chromium is generally subdivided into 'hard' and 'decorative' plating methods. Hard coatings (2 to 250 μm) are thicker than decorative coatings (<2μm) and are used for engineering application, often in conjunction with a sub-layer of nickel. Various types of hard chromium exist, including microcracked chromium, microporous chromium, porous chromium and crack-free chromium. The latter can only be produced in thicknesses up to 2.5 μm, whereas, the porous or cracked chromium deposits can be up to ~150 μm. Indeed, it is essential that the microdefective coatings have a minimum thickness ~80-120 μm in order to confer adequate corrosion resistance. Micro-cracked chromium has a Vickers hardness of 800 to 1000 kg/mm^2, while crack-free chromium has a Vickers hardness ~425 to 700 kg/mm^2. There is a lack of hardness data for microporous chromium and porous chromium. The former should not contain less than 15,000 pores per square centimeter; this favours a lateral corrosion path, delaying downward penetration. The formation of microporous chromium is achieved by a specialised plating method involving the use of suspended inert particles. Porous chromium plating is, however, developed by an etching procedure after electrodeposition of the chromium; this coating has a coarser distribution of pores. These are designed to retain lubricant, in sliding and bearing type applications. Chromium itself is a base metal; it is the external chromium oxide layer which provides the observed tolerance to atmospheric corrosion. For an excellent review see J. K. Dennis and T. E. Such: 'Nickel and Chromium Plating', 3rd ed; 1993, Cambridge, Woodhead Publishing.

cladding. Most commonly achieved by roll bonding two dissimilar metals together above their recrystallisation temperature (hot working), e.g., austenitic stainless steel can be clad to mild steel, producing corrosion resistant steel sheet at a fraction of the cost of solid stainless steel.

cleaning. Any method of removing surface contamination.

closed-field unbalanced magnetron sputter deposition. A recent innovation whereby at least two unbalanced magnetron sources are placed inside a vacuum chamber in such a manner that their external, unbalanced magnetic fields merge, forming an enclosed space (the closed-field configuration) in which electrons and ions are trapped; ions are prevented from reaching the chamber walls and their incidence of interception of the negatively biased substrates is significantly raised (diagram). Hence, even a small substrate electrical negative bias (~150 volts) can achieve a respectable current density (~5-10 mA/cm^2), sufficient to obtain temperatures ~400°C, even if pressures no longer sustain a DC glow discharge plasma at the substrate, e.g., at 3×10^{-3} torr. In fact, substrates are 'bathed' by plasma that extends out from the magnetron sources. The technique has been developed by Dr Bill Sproul, BIRL, Northwestern University, USA, and is now the basis of two commercial sputter systems, one deploying three unbalanced magnetrons, the other deploying four. Also see *unbalanced magnetron.* Further information is given in W. D. Spraul, P-J. Rudnik, M. E. Graham and S. L. Rohde, *Surface and Coatings Technology*, 1990, **43/44**, 270-278.

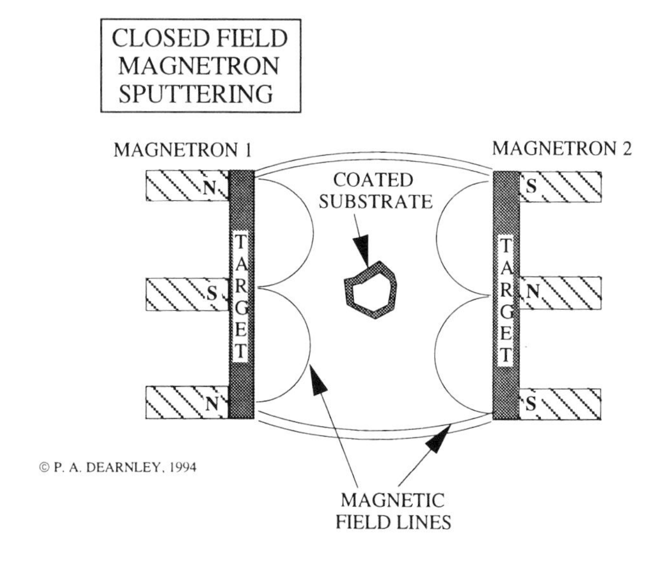

closed loop partial pressure control. A method of maintaining constant 'reactive gas' partial pressure during magnetron sputtering or ion plating. The appropriate mass number is "sensed" by a mass spectrometer and the voltage signal sent to a logic control unit which compares it against a prior set-point voltage. The same controller opens or closes a solenoid or piezo-electric actuated valve to achieve the desired partial pressure (See diagram). The technique is also useful for preventing *target poisoning* in magnetron sputtering.

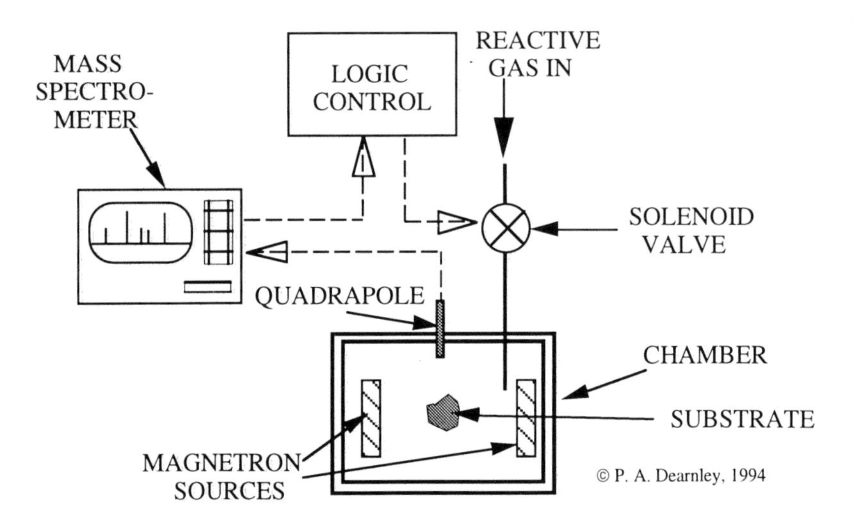

CO_2 laser. See *carbon dioxide laser.*

coating. (i) - verb - The action of creating a coating, e.g., by a plasma torch; (ii) - noun - A surface covering, usually of very different constitution to the substrate, which renders an improvement in corrosion or wear resistance or some other desired property.

coating compatability. The chemical compatibility of a coating with respect to its substrate. This quality is essential, particularly for high temperatures service, e.g., for coatings that provide oxidation protection for nickel alloy turbine blades.

coating hardness. The inherent hardness of a coating. For thin coatings, ~3-5μm, this is less easy to determine. Theoretical models exist, however, that enable the influence of the substrate to be taken into account. For example see O. Vingsbo, S. Hogmark, B. Jönsson, and A. Ingemarsson, in P.J. Blau and B.R. Lawn (eds): *Microindentation Techniques in Materials Science and Engineering*, 257-271; ASTM STP 889, 1986, Philadelphia, American Society for Testing and Materials. Yet another approach is to use the *nanoindentation hardness* technique.

coating interfacial shear strength. The interfacial shear strength along the coating substrate interface. Technically difficult to measure.

coating porosity. Holes or cavities in a coating, which may or may not be surface connected. Expressed as a volume or area fraction. It can be determined by image analysis of microsections or by density measurement methods (e.g., mercury infiltration).

coating thickness. The distance between the external coating surface and the coat/substrate interface, usually expressed in μm (10^{-6} m).

cobalt plating. An electroplating process in which a cobalt coating is deposited on the surface of an object. Although comparable to nickel plating, it is rarely used because of the high cost of cobalt.

coefficient of friction (μ). The ratio of the tangential (friction) force (F) to the normal force (W) acting on two *sliding or rolling* surfaces, i.e.,

$$\mu = F/W$$

A dimensionless constant which should be inedependent of W (i.e., F is proportional to W; this is the first law of friction) providing both materials are stressed below their yield strength and that seizure does not take place. Because of the different contacting geometries μ can be significantly lower in rolling contact than in sliding contact. Also see *Laws of Friction.*

cold solvent cleaning. Degreasing by immersion in an appropriate organic solvent, usually at or slightly above room temperature; sometimes in combination with mechanical agitation. Compare with *vapour degreasing.*

colour golds. These comprise a variety of gold based electrodeposits that enable novel colouration and corrosion protection of jewellery and domestic ware. The colour of electrodeposited silver-gold alloy coatings can be varied between white-gold and yellow-green, while electrodeposited copper-gold alloy coatings can vary from pink-gold, through rose-gold to red-gold. Cadmium-gold alloy electrodeposits are green.

combined thermochemical treatment. Any surface treatment involving two or more sequential thermochemical diffusion steps, e.g., see *multicomponent boriding.*

combustion flame spraying. Any thermal spray method that exploits the thermal and kinetic energy of combustion. See *detonation gun spraying high velocity air fuel (HVAF) spraying* and *high velocity oxygen fuel (HVOF) spraying.*

combustion wire gun spraying. A thermal spray process in which coating material in the form of a single wire or rod is passed into an oxyacetylene flame and melted. The liquid droplets so formed are instantaneously atomised by a blast of compressed air, propelling very fine molten droplets out of the gun as a spray. A wide variety of metals can be sprayed including molybdenum, austenitic stainless steels, aluminium bronze and nickel-aluminium. Al_2O_3 ceramic can also be used but the rod length is restricted to 250 mm. A recent development utilises ceramic powder encapsulted in plastic sleeving; the polymer decomposes during spraying and does not contaminate the ceramic coating. This form is available in longer lengths, up to 1.50 m. Also see *arc wire spraying.*

composite coating or layer. Any coating comprising (i) a mixture of finely dispersed metal and ceramic phases or (ii) a number of overlapping coatings of dissimilar materials that can be likened to an American sandwich. Also see *duplex surface engineering.*

compositionally modulated coating. See *modulated coating.*

compound layer.

> (i)'Outermost part of the case produced by thermochemical treatment, consisting of chemical compounds formed from one or more of the elements introduced by the treatment and elements of the base material' – IFHT DEFINITION

(ii)Also called white layer and (less commonly) nitrided layer. Refers to the outermost surface layer of nitrided or nitrocarburised steels comprising γ'-Fe_4N and/or ε-$Fe_{2\text{-}3}N$. Usually this zone should not exceed 10μm in thickness. Nitriding conditions are sometimes adjusted to completely avoid its formation (see *bright nitriding*). The wear resistant qualities of the ε-$Fe_{2\text{-}3}N$ phase are exploited in nitrocarburised plain carbon steels. See *Nitrocarburising, Nitrotec and Nitrotec S.* Contention still exists concerning the relative wear resistance of γ'-Fe_4N and ε-$Fe_{2\text{-}3}N$. Experience has shown that in examples where both phases coexist, detrimental residual stresses are built-up which contribute to exfoliation of the layer.

concentration overpotential. A change in potential of an electrode resulting from changes in the electrolyte composition (in close proximity to the electrode-electrolyte interface) caused by an electrode reaction.

concentration profile. Variation in concentration of elements as a function of depth below the surface. Various techniques are used to derive such profiles, e.g., SIMS, GDOES and SES. Also see *hardness profile.*

cone cracks. See *Hertzian cracks.*

contact fatigue. See *rolling contact fatigue.*

contact plating. Also termed galvanic contact plating. A form of electroless plating in which a coating of some metal M is produced on a metal object by immersion in an electrolyte enriched in M. In addition a solid piece of metal M is attached to the object to be coated. The action is galvanic in nature; it is a pre-requisite that M should be relatively more anodic than the object being coated.

contact stresses. The elastic and plastic stresses developed in a material during point contact loading. For example, the stresses developed in an object during indentation hardness testing. Also see *Hertzian stresses.*

continuous wave (CW) laser. Any laser capable of delivering a continuous beam of laser radiation. CO_2 lasers commonly operate in this mode. Also see *pulse mode laser.*

contour hardening. Also called shell hardening. Achieved with a steel of limited hardenability. Commonly, a cold work tool steel (~1.0 wt-% C), perhaps of 50 mm section size, is austenitised, quenched and tempered. Due to its low hardenability, the surface transfroms to martensite, while the core does not. This result in a tool with a tough core and a highly wear resistant surface.

conversion coating. A non-metallic and relatively complex, inorganic coating applied to metallic surfaces by immersion in an appropriate solution; the solution reacts with the metallic surface forming a stable compound layer. Common treatments include *phosphating* and *chromating*.

copperising.

'Diffusion metallising with copper' – IFHT DEFINITION.

The aim of copperising is to increase corrosion resistance and electrical conductivity of steel objects. The process is accompanied by a considerable increase in mass and dimensions of the object under treatment. See *minor thermochemical diffusion techniques.*

copper plating. Applied by electrodeposition, copper plating is often used in combination with, or as a substitute for, nickel undercoats prior to chromium plating. It has good levelling characteristics. It is also used in its own right for electrical contacts on printed circuit boards. The electrolyte used for copper plating is based on copper sulphate and sulphuric acid.

core.

(i)'Inner region of a heat treated object' – IFHT DEFINITION

(ii)The interior part of a metal object that is not directly modified by a diffusion or deposition treatment.

core hardness. The hardness of the substrate beneath a coating or diffusion zone. A term commonly used in the heat treatment sector in connection with nitrided or carburised steels.

corrodibility. A rarely used (and qualitative) term used to describe the susceptibility of a metal or alloy to corrosion attack. Also see *pitting index.*

corrosion. The degradation of a conducting surface through electrochemical action.

corrosion control. The deployment of techniques to control the corrosion rate of an object to an acceptable or ecenomic level.

corrosion fatigue. Fracture failure resulting from the conjoint action of cyclic surface stresses and a corrosive environment.

corrosion potential. Sometimes termed compromise potential or mixed potential. E_{corr} the potential resulting from mutual polarisation of the interfacial potentials of partial anodic and cathodic reactions that comprise the total corrosion reaction. Also see *bimetallic corrosion.*

corrosion product. Usually referring to the metallic reaction product resulting from a corrosion reaction, although equally applicable to ions or gases produced in the same way.

corrosion rate. The rate of progress of corrosion, often expressed as the rate of penetration, e.g., mm/year or as a weight loss per unit area per unit time, mg. dm^{-2} day^{-1}

corrosion resistance. The resistance of any surface to atmospheric or aqueous corrosion, i.e., resistance to degradation through electrochemical action.

corrosion-wear (or corrosive-wear). Wear taking place in a corrosive environment. The wear processes could involve abrasion or erosion or other mechanisms. Localised cold-working of the surface, e.g., caused during abrasion, renders the surface more prone to corrosion. Marine engine and food processing components are application sectors where corrosion-wear plays an important role in determining component servicability. The magnitude of the 'rest' and 'active' component cycles are of particular significance in this regard.

crevice corrosion. A localised corrosion taking place at crevices formed between two adjacent surfaces, at least one of which is metallic. Stainless steels are especially prone to crevice corrosion.

critical pitting potential. The most negative potential required to initiate surface pits, when a metal is held within the passive region of potentials.

curing. In connection with polymeric coatings. The process of obtaining a dry coating with fully developed properties. Thermosetting polymers generally require baking, while others can be cured by ultraviolet light.

curtain coating. A modification of the plastic flow coating process, involving the use of a continuous curtain of polymer solution flowing from an adjustable slot, beneath which a metal strip or sheet is passed at controlled speed.

CVD. See *chemical vapour deposition.*

CVI. See *chemical vapour infiltration.*

cyaniding

> 'Carbonitriding using a cyanide-based salt bath'
> – IFHT DEFINITION.
> See *carbonitriding.*

D

DC diode PAPVD. A plasma assisted PVD process in which the object to be coated is made one of the cathodes of a DC glow discharge, the chamber walls acting as the earthed anode.

DC triode PAPVD. See *triode ion plating* and *triode sputter deposition.*

deboriding treatment. A thermochemical method for the removal of boron from a prior borided steel surface. The process is conducted in an atmosphere containing H_2 and CH_4 at temperatures of 900-1020°C. Boron is usually removed in the form of B_2O_3.

decarburisation. Loss of carbon from a steel surface to a level well below that of the core, often as the result of prolonged high temperature exposure in an oxidising atmosphere.

decarburisation depth.

> 'Distance from the surface of a decarburised object to a specified limit which depends on the character of the decarburisation. The limit can be set in terms of a specified structure, hardness, or carbon content' – IFHT DEFINITION.

decorative chromium plating. See *chromium plating*

decorative coating. Any coating primarily produced to enhance the appearance of an object. For example see *gilding.*

degreasing. Any methods that achieve the removal of organic compounds, such as grease or oil, from the surface of an object, e.g., see *vapour degreasing.*

denitriding.

> 'Thermochemical treatment for the removal of excess nitrogen from the surface of a nitrided object' – IFHT DEFINITION.

deposit. A synonym for coating (noun), but also referring to any adherent surface solid formed, for example, as a result of wear or corrosion.

deposition. The act of laying down a metallic or non-metallic layer onto a substrate. There are numerous deposition techniques; these include electroplating, plasma spraying, PVD and CVD.

deposition rate. Used in connection with coating technologies, like plasma spraying, plasma assisted PVD or CVD. Expressed in mm/min, μm/min or μm/hr.

derusting. See *descaling.*

descaling. Any process used for the purpose of removing the scale from heavily oxidised metal. This can be achieved by mechanical (abrasive blasting, tumbling, brushing) or chemical means (alkaline descaling, acid descaling (pickling) or salt bath descaling).

designer surface. An engineered surface with optimal constitution, treatment depth and properties, designed with intent for specific application(s).

detonation gun (D-gun) spraying. A combustion spraying technique whereby particles of coating material, typically WC-Co, are passed through a combustion tube, fuelled by a mixture of oxygen and acetylene gases. Ignition is provided by a spark plug 4.3 or 8.6 times per second. The composition of the combusting gas can be adjusted to be oxidising, neutral or reducing. The flame temperature is ~3000 K and coating powder particles can attain velocities ~ 800 m/s. It is an extremely noisy torch (150 dB) requiring intensive screening and operator protection. Union Carbide Corporation is the principal owner and operator of this technology. Also see *high velocity air fuel (HVAF) spraying* and *high velocity oxygen fuel (HVOF) spraying.*

D-gun spraying. See *detonation gun (D-gun) spraying.*

diamond coatings. Diamond coatings can be deposited by many methods. Presently, the most favoured technique is microwave plasma CVD. Diamond coatings are crystalline and are usually characterised by a combination of methods which include, Raman spectroscopy, X-ray diffraction and surface morphology. Also see *hot filament CVD* and *microwave plasma CVD.*

diamond-like coatings (DLC). Very hard (~3000–4000 kg/mm^2) carbon coatings with little or no crystallinity (amorphous). Initially such coatings proved unsatisfactory for tribological protection because of poor substrate adhesion and unfavourable residual stress. Some recent progress has been made whereby appropriate sub-layers of transition metal carbides and nitrides are deposited prior to deposition of a DLC. This is particularly important when depositing DLC onto metallic substrates. DLC's can be deposited by sputter deposition or plasma assisted CVD.

diffuse cycle. The second part of the boost-diffuse method. See *vacuum carburising* and *plasma carburising.*

diffusion coating. Any process that produces a surface enriched in another element through solid state diffusion and (usually) resulting in the formation of intermetallic or interstitial compounds. Processes include *aluminising, boriding, chromising, sherardising, siliconising* and *vanadising*. By convention, it specifically *excludes* carburising, carbonitriding, nitriding, and nitrocarburising.

diffusion metallising.

> 'Thermochemical treatment involving the enrichment of the surface layer of an object with one or more metallic element' – IFHT DEFINITION.

Diffusion coating with metals alone. Processes include *aluminising, chromising, sherardising* and *vanadising* but should exclude the technology of siliconising, since silicon is a non-metal. However, some authors still regard siliconising as a diffusion metallising process.

diffusion period. That time of a thermochemical treatment when the active specie(s) diffuse into the substrate. To provide distinction from, for example, the heating and cooling cycles.

diffusion profile. The concentration of elements in the diffusion zone following a thermochemical diffusion treatment. Also see *concentration profile*.

diffusion wear. A term first invoked by Trent and Loladze in the 1950s, to account for the rake face cratering of cemented carbide cutting tools, observed after cutting plain carbon and low alloy steels at relatively high speed (>100m/min). Other workers have termed this effect 'dissolution/diffusion wear' or 'solution wear'. For cemented carbide tools, comprising WC–Co or WC–(W,Ti,Ta,Nb)C–Co (termed 'straight grade' and 'steel cutting grade' carbides respectively), the WC grains become smoothly worn. In the case of 'steel cutting grade' cemented carbides, microsections, made normal to the rake face surface, reveal that the WC phase is preferentially worn; the (W,Ti,Ta,Nb)C phase being less severely worn, stands proud of the adjacent WC grains. The preferential wear of the WC phase is attributed to its higher solid solubility limit in γ-Fe (present in the hot steel chip shear zone during cutting) compared to that of the (W,Ti,Ta,Nb)C phase.

Other tool materials, like high speed steel and ceramics can be worn by this type of mechanism, e.g., when sialon tools are used to cut low alloy and plain carbon steels at cutting speeds above 200 m/min; similarly Al_2O_3-ZrO_2 ceramics exhibit a similar effect when used to cut titanium alloys at cutting speeds above 50 m/min. TiN and TiC coatings, deposited by PVD or CVD, are also worn in this way when used to cut steels above 100 m/min although other mechanisms like *discrete plastic deformation* make a significant contribution. For further details see *Metal Cutting* by E.M. Trent, Butterworths, 3rd Edition, 1991.

diffusion zone.

> 'That part of the case where elements introduced by thermochemical treatment are held in solid solution or partially precipitated in the matrix. The concentration of these elements decreases continuously towards the core' – IFHT DEFINITION

A term used especially in connection with nitrided steels. In this example it refers to the zone of coherent nitride precipitation, residing beneath the compound layer, which constitutes 95 to 99% of the depth of the nitrided 'case', see *nitriding*. The nitrogen content of the diffusion zone ranges from nearly 20at-% adjacent the compound layer to below 1at-% adjacent the core structure.

dimensionless wear coefficient. See *Archard's wear equation.*

diode sputter deposition. A sputter deposition process in which the substrate serves as a positive or positive-ground electrode, relative to the cathodic sputter sources; a less commonly used configuration. Also see *bias sputter deposition.*

dip coating. Any technique whereby a coating is formed by immersing a solid in a liquid which subsequently bonds to the solid surface. The most common types include *hot dip aluminising* or *hot dip galvanising*. The term also refers to coating by immersion in ceramic slip or even polymeric solutions (organic paints).

dip coating with ceramics. See *ceramic dip coating.*

dip coating with plastics. See *plastic dip coating.*

dip painting. See *plastic dip coating.*

direct current plasma CVD. Plasma assisted CVD in which the plasma is generated by a direct current (DC) power supply.Often more problematic than *radio frequency (RF) plasma CVD.*

discrete plastic deformation. A common wear mechanism of PVD and CVD ceramic coatings (TiC, TiN and Al_2O_3) when used to cut steel at speeds above 100 m/min. The mechanism increases with increasing cutting speed (temperature) and is the principal rake face wear mechanism of Al_2O_3 coated cemented carbide tools. The mechanism culminates in the removal of wear fragments through ductle fracture (see diagram, sequence 1 to 4 and micrograph). It should be appreciated that coating temperatures often exceed 1100°C when cutting steel; at these temperatures TiC, TiN and Al_2O_3 behave plastically. Also see P. A. Dearnley, R. F. Fowle, N. M. Corbett and D. Doyle, *Surface Engineering*, 1993, **9**, 312-318.

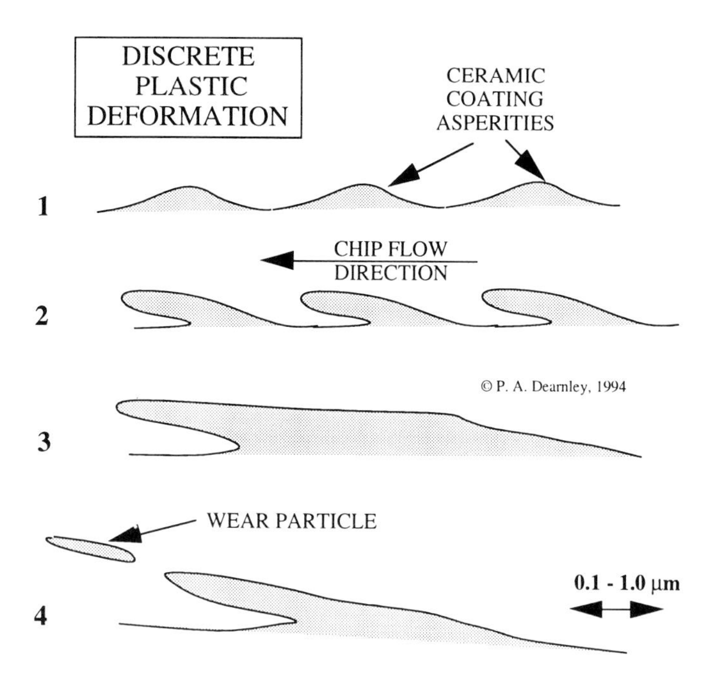

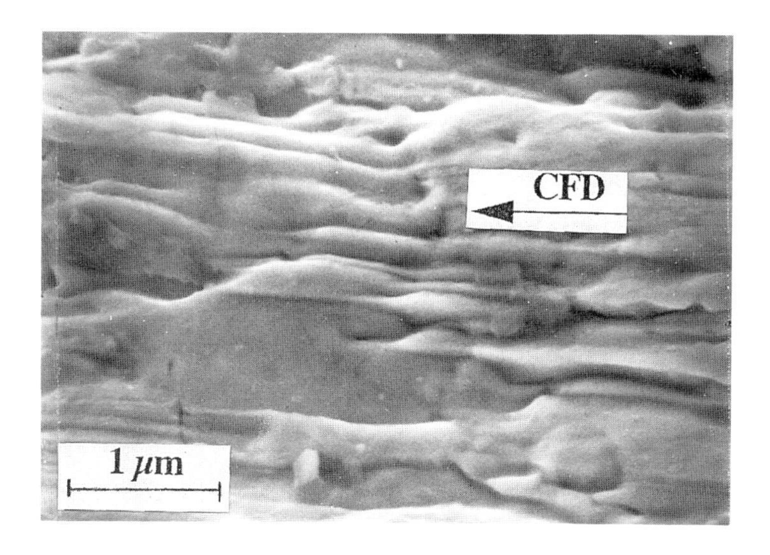

discrete plastic deformation. *A CVD–TiC coating undergoing rake face wear through discrete plastic deformation caused by the high temperatures and shear stresses developed while cutting an AISI 1046 normalised steel at 300 m/min 0.25 mm/rev. CFD = chip flow direction. Field emission SEM.*

displacement CVD reaction. Also known as an exchange reaction. A CVD reaction between a volatile compound (MX_2) and a substrate (Z):

$$MX_{2\ (g)} + Z_{\ (s)} \rightarrow M_{\ (s)} + ZX_{2\ (g)}$$

This reaction is usually self-limiting i.e., it stops after a solid layer, $M_{(s)}$, has been deposited which serves as a diffusion barrier to further reaction. For example see *gaseous chromising.*

DLC. See *diamond-like coatings*

double-stage nitriding. See comments on Floe Process under *gaseous nitriding.*

drip-feed carburising. See *gaseous carburising*.

dry blast cleaning. A general term encompassing shot blasting, grit blasting and sand blasting.

dry galvanising. Hot dip galvanising in which the object, after degreasing and pickling, is immersed in an aqueous flux solution, then dried and immersed in a molten zinc bath. A rather inappropriate term.

drying. The process of removing residues from washing operations by means of hot or cold air, freon, infra-red radiation or centrifugal action.

dull finish. A surface of poor specular reflectivity. *Matt surface* is also an acceptable term.

duplex nickel plating. A nickel electroplating process involving the deposition of two layers of nickel: (i) semibright and; (ii) bright. These serve as an undercoat for subsequent chrome plating.

duplex surface engineering. A combination of two sequential (and mutually compatible, from the treatment temperature perspective) surface engineering methods to achieve a synergistic surface property design. At the time of writing, much interest is being shown in extending the application range of thin (~5μm) ceramic coatings (like TiN and TiC) to low yield strength (~500-1000M.Pa) substrates by prior strengthening via a thermochemical diffusion treatment. Feasible duplex treatments include: (i) CVD after carburising, followed by quenching (to permit the γ-Fe —> α'-Fe transfomation); (ii) PVD after nitriding. Note: carburising temperatures ~ CVD temperatures; PVD temperatures ≈ nitriding temperatures. Initial trials have been carried out in two distinct operations. More recent efforts have focused on enabling each process to be carried out sequentially in the same vessel.

durability. The ability of a surface engineered material to endure in-service wear, corrosion or fatigue.

E

EBPVD. See *electron beam evaporative PVD.*

EDS. Energy dispersive spectrometry; the same as EDX.

EDX. Energy dispersive analysis by X-rays. An electron beam (~20-30 keV) is focused onto a the sample surface, which must be electrically conducting; among the many surface effects, X-rays, of characteristic energy are produced. An optimal sample-electron beam distance is used to assure X-rays of sufficient intensity are brought to focus at a solid-state detector, chilled to liquid nitrogen temperature. The detector is frequently isolated from the microscope vacuum chamber by a Beryllium window. X-rays with an energy in the range of 1 to 20 keV are efficiently detected allowing the simultaneous detection of all elements heavier than boron, using at least one of the principal K, L, or M emission lines. Some modern devices have ultra-thin windows or offer windowless detectors allowing the detection of elements down to Li. However, it should be appreciated that the detection of low atomic number elements below oxygen is accompanied by a poor signal-to-noise ratio, making even semi-quantitative analysis difficult. Quantification of heavier elements is easier; nowadays this is achieved routinely by dedicated computer software. The method is very rapid. EDX is an excellent tool for gaining an immediate idea of chemical composition. Elemental mapping, in conjunction with secondary electron imaging is also a useful feature of this method. For high accuracy chemical composition analysis, however, EDS is inferior to WDX. The latter method is significantly better for light element analysis.

effective case depth after carburising

> 'Distance between the surface of a carburised object and the region where the Vickers hardness under a load of 9.81 N is HV1 = 550'
> – IFHT DEFINITION.

Also see *case depth.*

effective depth of hardening.

> 'Perpendicular distance between the surface of a quench-hardened object and a limit defined by a specific hardness value or microstructure'
> – IFHT DEFINITION.

effective nitrided case depth.

'Nitrided case depth determined on the basis of a specified hardness value' – IFHT DEFINITION.

Also see *case depth.*

EHL. See *elastohydrodynamic lubrication*

elastohydrodynamic lubrication (EHL). Lubrication between two counterformal (non-conforming) surfaces where very high contact pressures arise, e.g., at the point contacts made between intermeshing gear teeth.

electric arc spraying. See *arc wire spraying.*

electrobrightening. See *electropolishing.*

electrochemical cleaning. Also termed electrocleaning. See *cathodic cleaning, electropolishing* and *periodic reverse electrocleaning.*

electrocoating with plastics. See *electrophoretic coating, electrostatic fluidised-bed coating* and *electrostatically sprayed plastic coatings*

electrodegreasing. See *cathodic cleaning, electropolishing* and *periodic reverse electrocleaning.*

electrodeposit. Any coating formed by electrodeposition.

electrodeposition. The techniques of forming coatings via *electroplating* or *electrophoresis*. Also see *electrophoretic coating.*

electrodeposition of plastics. See *electrophoretic coating, electrostatic fluidised-bed coating* and *electrostatically sprayed plastic coatings*

electrode potential (*E*). The difference in electrical potential between a metallic electrode and its contacting electrolyte. Usually expressed in relation to the standard hydrogen electrode (S.H.E.). Its magnitude is equal to the e.m.f. generated by a cell comprising the same metal and a S.H.E. If the metal electrode becomes cathodic its electrode potential is positive, conversely if it becomes anodic its electrode potential is negative. If the species are in their standard state, the electrode potential (E) is then known as the *STANDARD ELECTRODE POTENTIAL* (E^{ϕ}) Also see *e.m.f. series.*

electrogalvanising. See *zinc plating.*

electroless nickel plating. A nickel plating process that does not require the action of an externally supplied electric current. Instead, an autocatalytic chemical reduction of nickel ions is acheived by the application of hydrophosphite, aminoborane or borohydride compounds. Depending upon the reducant, the coatings contain nickel phosphide or nickel boride in a nickel matrix. The Vickers hardness of the as deposited coating is ~500–550 kg/mm^2. After heat treating at 1000°C for 1 hr this increases to ~ 900–1000 kg/mm^2; longer treatments at lower tempertature can also be used to increase hardness, e.g., 230°C for 28 h. Such coatings have good wear *and* corrosion resistance.

electrolysis. Decomposition of a liquid and/or solid by the action of an electric current. Electrolysis constitutes the basis of electroplating.

electrolyte. Any liquid capable of conducting an electrical current. Two surface engineering methods use electrolytes: (i) electroplating, e.g., nickel plating; (ii) electrolytic salt bath thermochemical processes, e.g., electrolytic boriding. In the first case the electrolyte comprisies ionic salts dissolved in water, in the second, fused ionic salts are used.

electrolytic boriding.

> 'Boriding by electrolysis in a molten borax-based medium' – IFHT DEFINITION.
>
> See *salt bath boriding.*

electrolytic borochromising. A liquid phase multicomponent boriding treatment for the simultaneous diffusion of boron and chromuium into steel surfaces. It is carried out electrolytically in a molten salt mixture containing 90–95wt-% $Na_2B_4O_7$ (borax) and 5–10wt-% Cr_2O_3, or 75-80wt- % B_2O_3, 12–22wt-% NaF and 3-8wt-% Cr_2O_3. The process is conducted at a current density of 0.1–0.2 A/cm^2 at temperatures of 800–1000°C for 1–6 hours. The thickness of the layer produced does not exceed 0.1mm. This treatment is less effective than those involving sequential diffusion of boron followed by chromium. Also see *multicomponent boriding.*

electrolytic carburising.

> 'Carburising involving the passage of current between the object to be treated (which acts as a cathode) and a graphite anode, usually in a molten salt bath' – IFHT DEFINITION.

Carburising in a molten salt mixture containing alkaline, alkaline–earth metal carbonates and halides. The object being treated acts as a cathode, while the anode comprises a graphite or carborundum rod. The process is carried out at 950°C. Rarely practised outside Eastern Europe. Also see *salt bath carburising.*

electrolytic cleaning. See *cathodic cleaning*, *electropolishing* and *periodic reverse electrocleaning*.

electrolytic degreasing. See *cathodic cleaning*, *electropolishing* and *periodic reverse electrocleaning*.

electrolytic phosphorising. Liquid phase phosphorising carried out by the electrolytic method in a molten salt mixture containing (by weight) 30% NaCl and 70% Na_3PO_4. The process is conducted at a current density of 0.4–0.5 A/cm^2 at temperatures of 920–940°C for 3–4 hours. The thickness of the layer produced is about 0.1mm.

electrolytic polishing. See *electropoloishing.*

electrolytic salt bath descaling. Salt bath descaling in which the molten salt acts as an electrolyte and the object to be descaled may be either cathodically or anodically polarised. This process is mainly used for high-alloy tool steels. Rarely practised outside Eastern Europe.

electron assisted PVD. See *supported glow discharge plasma.*

electron beam. A high power density beam generated by a device termed an electron gun. This comprises a pair of anular electrodes and a resistively heated tungsten filament which emits electrons via thermionic emission. The electrons are accelerated by respectively passing them through the centres of an anular cathode and anode. After exiting the gun, the electrons are focused by an electromagnetic field, generated by a solenoid. Very high accelerating voltages can be used (~500 keV) and power densities of 10^6 to 10^7 W/cm^2 are easily achieved. Higher power densities are achievable by operating in pulse-mode; a similar principle to that deployed in Nd–YAG and excimer lasers. In surface engineering an electron beam is complimentary to a laser. It is used as an evaporation device, e.g., in ion plating, or for surface alloying and transformation hardening of metallic materials. Energy coupling to metallic surfaces is outstanding, but electron beams have the disadvantage of not been applicable to non-conducting surfaces, like dielectric ceramics (e.g., Al_2O_3). Despite development devices that have demonstrated the feasibility of operation at atmospheric pressure (e.g, see Y. Arata: *Plasma, Electron & Laser Beam Technology,* 1986, Ohio, American Society for Metals), the routine industrial use of electron beams remains restricted to the confines of a vacuum chamber (minimum pressure $\approx 10^{-4}$ torr).

electron beam alloying. See *power beam surface alloying.*

electron beam amorphisation. See *electron beam glazing.*

electron beam carburising. Carburising involving electron beam heating of surfaces into the austenitic range. The carbon can be supplied as a gaseous hydrocarbon, but more readily as a pre-placed graphitic surface coating. During electron beam heating, the carbon dissolves into the austenite and the structure is allowed to self-quench by conduction into the relatively cool core. Treatment depths are relatively shallow <50 μm. The technique is not widely used.

electron beam cladding. An electron beam treatment applied to a metallic surface which is momentarily surface melted to enable the encapsulation of preplaced or injected ceramic powder particles. The treatment is applicable to ferrous or non-ferrous metals and alloys. There has been a lot of interest in recent years in applying such treatments to aluminium and titanium alloys because both suffer from the disadvantage of responding poorly to conventional thermochemical diffusion treatments, i.e., they develop very shallow treatments, if any. Electron beam cladding offers the possibility to develop deeply hardened surfaces, up to 1 mm. The process is analogous to *laser cladding.*

electron beam evaporative PVD. An ion plating or vacuum evaporation process that uses an electron beam (eb) to vaporise the metallic source constituent(s) of a coating. It is a common requirement of such devices that the electron beam should be housed in a differentially pumped vacuum chamber attached to the main vessel. This enables the electron beam to operate at the preferred pressure of 10^{-4} torr while allowing a glow discharge plasma to be created at higher pressure, ~10^{-2} torr, in the main deposition vessel (diagram). The electron beam is steered into the main chamber, via a small aperture, using a system of electromengnets. Sufficient power density is available to vaporise even refractory metals like tungsten and molybdenum. Also see *ion plating.*

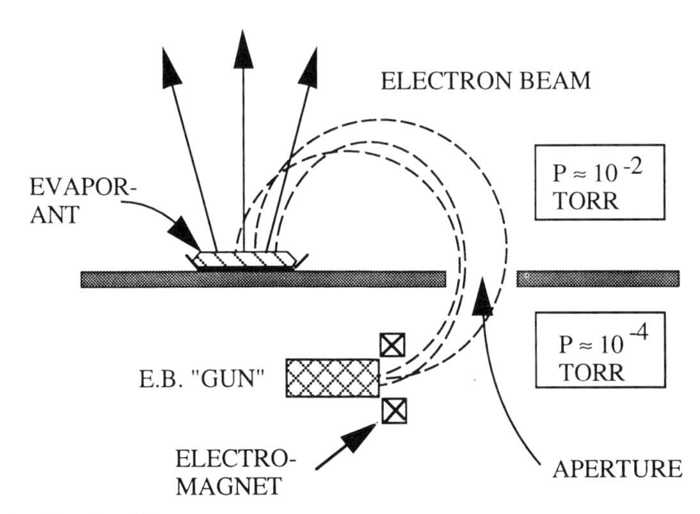

electron beam gas alloying. See *power beam surface alloying.*

electron beam glazing. An electron beam treatment aimed at producing an amorphous/glassy surface layer. It involves electron beam heating a surface with power densities ~ 10^5 to 10^7 W/cm^2 and interaction times ~ 10^{-4} to 10^{-7} seconds; subsequently the surface rapidly solidifies at cooling rates exceeding 10^5 K/s, which suppresses the nucleation and crystallisation processes, thereby promoting vitrification/amorphorisation. Less widely practised than *laser glazing*.

electron beam gun. See *electron beam.*

electron beam hardening.

> 'Hardening using an electron beam as the heat source'
> – IFHT DEFINITION.
>
> See *electron beam transformation hardening.*

electron beam source. See *electron beam.*

electron beam surface alloying. See *power beam surface alloying.*

electron beam transformation hardening. A process involving electron beam heating of a steel surface into the austenitic temperature range followed by rapid self-quenching via conduction into the relatively cool core. The process produces relatively low distortion. It is claimed to be more economic than laser transformation hardening.

electron beam treatment. Any heat treatment utilising an electron beam as the principal heat source.

electron beam vitrification. See *electron beam glazing.*

electron probe microanalysis (EPMA). See *WDX*

electrophoresis. See *electrophoretic coating.*

electrophoretic coating. Charged colloidal particles of the material (a metal, ceramic or polymer) to be deposited are suspended in appropriate solution (carrier system), sometimes based on methyl alcohol and water. The particles are attracted towards the cathodically charged objects and become neutralised. It is usual practice to carry out a post deposition densification step. For example, aluminium particles deposited onto steel sheet can be densified by: (i) passing the material through compacting rolls and; (ii) sintering at temperatures ≈400-600°C. Parallel techniques exist for the deposition of ceramic powders like TiC. In such cases, densification is reliant on pressure sintering techniques at temperatures >1000°C; in the example of polymers, fusion is achieved by heating to temperatures ~150°C. The latter is also termed electrophoretic painting.

electrophoretic painting. A method of producing a polymeric coating. See *electrophoretic coating*

electroplating. The electrodeposition of thin adherent layers of metals or alloys onto a metallic substrate, thereby creating a composite material whose surface properties are dominated by those of the coating. Chromium and nickel plating are probably the two commonest forms of electroplating. Developments in electroplating technology include: pulse plating, modulated current plating, superimposed current plating and periodic reverse current electroplating.

electropolishing. Surface finishing of metallic objects by making them anodic in an appropriate electrolyte, whereby high points are preferentially dissolved, resulting in a bright and level surface with high specular reflectivity. In a sense the process can be likened to electroplating in reverse, i.e., the articles are anodic, while in electroplating they are cathodic. Hence, the process is sometimes called reverse current cleaning or anodic cleaning. A wide variety of metals and alloys can be electropolished, they include: aluminium, stainless steels, brass, copper and nickel-silver. Polishing solutions are genereally based on phosphoric acid and are used at a current density ~15 to 80 amp/dm^2.

electrostatic fluidised-bed coating. A fluidised-bed method of coating parts with plastic/polymer. A d.c. high voltage electrode is used to electrostatically charge polymer powder particles which become attracted to the electrically grounded surfaces of the objects being treated. Following deposition, the objects are heat treated in an oven; this is set to a temperature sufficient to cause melting of the polymer and simultaneously enable good bonding to the metallic surface beneath. Before coating steel parts, it is important to vapour degrease and grit blast the component surfaces, followed by phosphating or chromating. Also see *fluidised-bed coating with plastics.*

electrostatically sprayed plastic coatings. A method of spray depositing polymeric coatings onto metals, especially steels. A fine polymer powder is fed from a hopper into an air stream and passed through a hand held spray pistol where it becomes positively charged using a voltage ~20-60 kV. The powder is projected out of the pistol as a fine aerosol. The powder particles are electrostatically attracted to the workpiece surfaces which are electrically grounded. Overspray is collected by a cyclone or forced extraction system containing filters. Following deposition, the objects are heat treated in an oven; this is set to a temperature sufficient to cause melting of the polymer and simultaneously enable good bonding to the metallic surface beneath. This method is a good technique for producing relatively thin polymer coatings (~50 to 200 μm), but capital cost of the equipment is quite high. Nylon is commonly applied by this method. Before coating steel parts, it is important to vapour degrease and grit blast the component surfaces, followed by *phosphating* or *chromating*. Also see *flame spraying of plastic coatings.*

e.m.f. series. Tabulation of standard equilibrium electrode potentials, relative to the standard hydrogen electrode, for electrode reactions of the general type:

$$M^{z+} (aq.) + Ze = M$$

The table is arranged in accordance with the sign and magnitude of the potential. Noble metals, like gold and platinum have positive standard electrode potentials and appear at the top of the series table; base metals like zinc and aluminium have negative standard electrode potentials and appear at the bottom of the series table.

emissivity. The ratio of the total emissive power density (W/cm^2) of a heated object to the emissive power density of a perfect black body radiator at the same temperature. The nearer to unity, the more perfect the emissivity. A dimensionless quantity. Symbol ε.

enamel. See *vitreous enamel coating.*

enamelability. Suitability of a given substrate material for vitreous enamel coating. It refers especially to steel substrates in which the carbon content should be as low as possible or in which carbon should be stabilised by the addition of titanium or niobium to prevent gas evolution caused by possible reaction between carbon and the oxides contained in the enamel. Also see *viteous enamel coating.*

enamelling. The procedure of applying a vitreous enamel coating. See *vitreous enamel coating.*

endothermic atmosphere.

'A controlled atmosphere produced by the catalytic decomposition of a hydrocarbon gas in a heated chamber' – IFHT DEFINITION.
Also see, for example, *gas carburising.*

energetic ions. Ions with high kinetic energy, acquired, for example, by being accelerated in an electric field.

energiser. See *activator.*

energy beam surface alloying. See *power beam surface alloying.*

engineering chromium plating. See *chromium plating.*

enhanced glow discharge. See *supported glow discharge.*

environmental impact of surface engineering. The following processes can be regarded as non-invasive on the environment:- plasma nitriding, PVD of transition metal nitrides and carbides, plasma spraying (APS and VPS) and nitrotec. Meticulous safety precautions and environmental safe guards are required when operating most salt bath methods, CVD, boronising and gaseous carburising processes.

epitaxial coating. See *epitaxy.*

epitaxial growth. The maintenance of epitaxy during coating deposition (growth).

epitaxy. Most commonly observed for coated single crystals (like silicon wafers), where a definite crystallographic orientation relationship exists between the coating and substrate. Epitaxy can be subdivided into homoepitaxy and heteroepitaxy. Homoepitaxy refers to the situation where the coating and substrate are made of the same material; heteroepitaxy refers to cases where the coating and substrate are made of dissimilar materials.

EPMA. See *WDX*

epoxy coating. A plastic coat containing thermosetting epoxide resins. They have excellent adhesion and chemical resistance.

erodent. The particulate material responsible for causing erosion.

erosion. A form of wear produced by the action of hard particles, suspended in a fluid (gaseous or liquid), repeatedly impacting on a solid surface. Particle velocities are usually in the range of 5 to 500 $m.s^{-1}$. Also termed slurry erosion, when the fluid is liquid.

erosive wear. Wear taking place by the mechanism of erosion. Commonly found in soil engaging tools, hydro-electric plant, beach-landing marine craft and aircraft components operating in desert environments

erosive wear resistance. The ability of a material to withstand erosion.

ESCA. Electron Spectroscopy for Chemical Analysis. An umbrella term for the high vacuum surface analysis equipment configured for both XPS and AES techniques.

eta phase (η-phase) zone. A layer or zone of η-phase formed by the decarburisation of cemented carbide (WC–Co or WC–[Ti,Ta,Nb,W]C–Co) susbstrates during chemical vapour deposition (CVD) of TiC, Ti(C,N) or TiN. η-phase is an M_6C type carbide, typically Co_3W_3C, formed at the coat-susbstrate interface and most frequently observed around the edges of cemented carbide objects (see micrograph) where carbon loss is greatest. With the advent of better carbon control: (i) during initial sintering of the cemented carbide substrates and; (ii) during CVD, the formation of η-phase has now become a less frequent problem. η-phase is embrittling and can lead to premature coating loss.

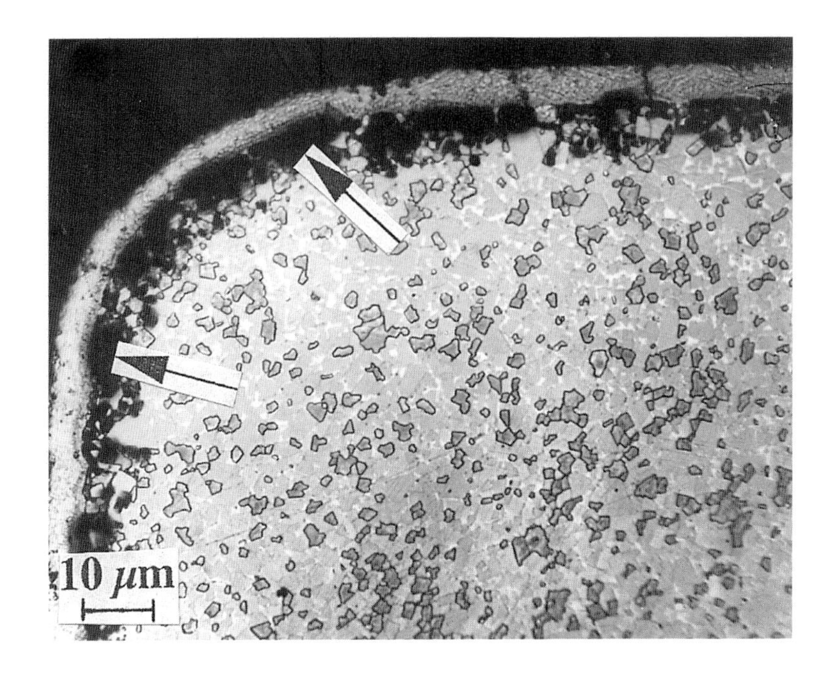

eta phase (η-phase) zone. Section prepared through the unused cutting edge of a CVD–TiC coated cemented carbide showing the development of an η–phase zone at the coat–substrate interface (arrowed). Light optical micrograph; specimen etched in 10% alkaline potassium ferricyanide. From P.A.Dearnley, Heat Treatment of Metals, *1987, (4), 83-91.*

etching. (i) a shallow chemical cleaning aimed at removing surface oxides or passive films by immersion in an appropriate acid or alkaline solution; (ii) a glow discharge cleaning process for the removal of passive oxide layers; usually achieved more efficiently with an RF rather than a DC plasma. Plasma etching is frequently a preparatory step before plasma assisted PVD.

evaporation coating. See *vacuum coating.*

evaporation source. A device, such as an electron beam or electric arc, used to vaporise the solid source component of a coating in evaporative source PVD.

evaporative source PVD. A PVD process in which vaporisation of the coating material is achieved by heating with resistive, electron beam, laser beam or electric arc devices. Deposition rates can be greater than 10 μm/min for pure metal coatings. Lower deposition rates are usual for metallic alloy or ceramic coatings. Also see *plasma assisted PVD.* A Tin coated high speed steel produced by this method is shown the micrograph. Also see *triode ion plating* and *ion plating.*

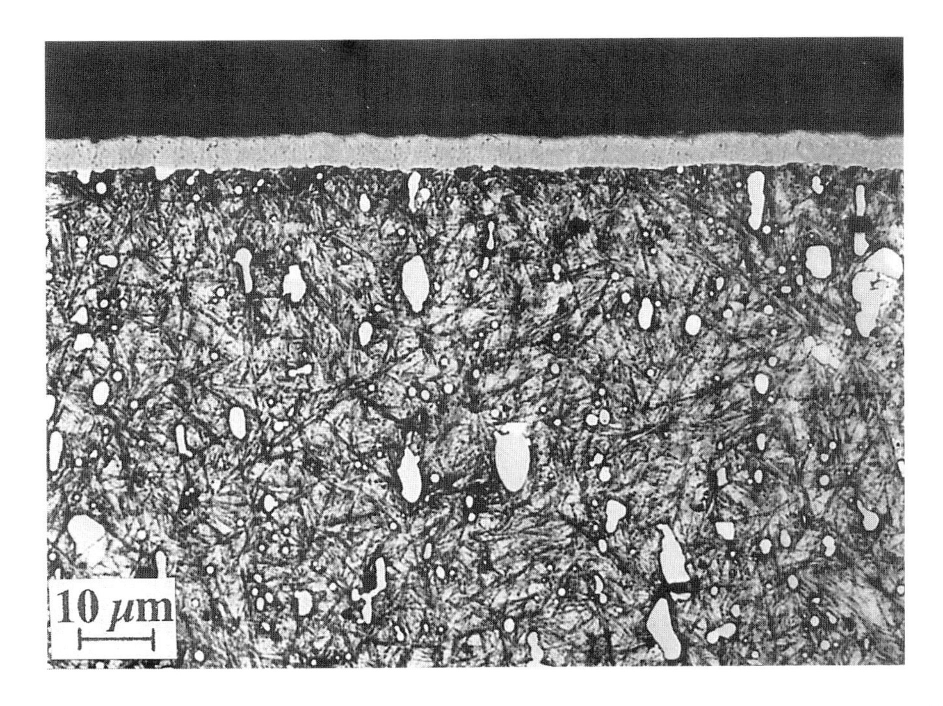

evaporative source PVD. *Normal microsection of an M42 high-speed steel cutting tool coated with TiN by an industrial evaporative (arc) source PVD system. Light optical micrograph (etched in 5% nital). From P.A.Dearnley,unpublished research, 1985.*

excimer laser. A gas laser in which the active medium is composed of gases like XeCl, XeF, ArF or KrF which irradiate light in the ultraviolate spectral range between 193 and 351 nm. They have a short pulse duration (11–40 ns) and a high pulse energy (1–4 J), which means they are ideally suited for the melting or evaporation of very thin surface layers (~ 1 μm/pulse). Hence, excimer lasers have been applied for the laser treatment of electronic devices and have shown promise for improving the bulk properties of ceramics (see *laser glazing*).

exfoliation. See *spalling.*

exothermic atmosphere.

> 'A controlled atmosphere produced by incomplete combustion of a hydrocarbon gas without heat supply' – IFHT DEFINITION.

expanded austenite. See *S-phase.*

explosive flame spraying. See *detonation gun spraying.*

F

facing by welding. See *weld hardfacing.*

fatigue strength. See *rolling contact fatigue.*

fatigue wear. See *rolling contact wear* and *rolling contact fatigue.*

fatigue wear resistance. The ability of a material to withstand rolling contact wear.

filler metal. The metal which is added in order to produce a deposited layer in the process of weld hard facing.

film. A term widely used in the electronics industries relating to thin (usually <1 μm) coatings applied to so called thin-film devices. Such coatings are frequently insulating or semiconducting in their electrical character.

film former. A component of polymeric coatings. See *binder.*

film thickness. See *coating thickness.*

filtered arc evaporator. An evaporation source designed to avoid the production of macroparticles. This can be achieved by a magnetic plasma duct. For a comprehensive review see P. J. Martin, *Surface Engineering,* 1993, **9**, (1), 51–57. Also see *arc source PVD.*

finish. See *surface finish*

finishing. See *surface finishing.*

finish coating. Sometimes referring to the outermost layer of a multilayered coating.

finite element method (F.E.M). A well established computational technique for determining the distribution of stresses in engineering structures under load. FEM has found recent use in surface engineering for the prediction of elastic stresses in low modulus metals coated with high modulus ceramic coatings during point contact loading.

flame hardening.

> 'Quench-hardening treatment involving flame heating, generally used for localised or surface hardening' – IFHT DEFINITION.

Sometimes called shorterizing. A convenient method of transformation hardening, requiring less capital investment than induction hardening. Oxyacetylene or oxygen-propane gas torch(es) provide the required heating to raise the surface temperature of medium carbon steels (≈0.3–0.5wt-%C) into the austenitic range; susbsequently the heated parts are quenched by spraying with water jets or by immersion in a quench tank, thereby developing the required martensitic case. There are three main types of flame hardening: (i) manual hardening comprises heating the objects to be hardened with a hand held torch. This method is useful for hardening small surface areas and small numbers of parts; (ii) spin hardening is employed for components with rotational symmetry; here the torches are mechanically fixed while the components are rotated on a motorised platform-fixture; (iii) progressive hardening is used for hardening flat surfaces or long sections, like machine slide-ways. The torch is passed over the surface, or the part is passed under a fixed torch, at speeds ≈50 to 200 mm/min. Immediately located behind the torch is a cold water jet which provides the quench medium. Also see *induction hardening.*

flame heating. Any heating operation utilising an oxyacetylene or oxygen-fuel gas torch.

flame spraying. See *thermospraying.*

flame spraying of plastic coatings. Plastic powder (particle size ~80–200 μm) contained in a hopper or fluidised bed is passed into a compressed air stream and fed into a combustion torch. Depending upon the torch design, powder is either passed through or around the combustion flame, but in either case the torch significantly differs from those deployed for metal spraying. The technique requires a high level of operator skill and is widely applied to components used in marine applications. Before coating steel parts, it is important to vapour degrease and grit blast their surfaces, followed, where possible, by phosphating or chromating. Also see *electrostatically sprayed plastic coatings.*

flame tempering. Surface tempering using an oxyacetylene flame. An unsatisfactory procedure, greatly inferior to furnace tempering.

flexibility. See *formability.*

Floe process. See *gaseous nitriding.*

flow coating with plastics. See *plastic flow coating.*

fluidised-bed boriding. See *fluidised-bed thermochemical diffusion methods*

fluidised-bed carburising.

> 'Carburising carried out in a medium of solid particles suspended in a flow of gas' – IFHT DEFINITION.
>
> See *fluidised-bed thermochemical diffusion methods*

fluidised-bed coating with plastics. A relatively low cost method of applying polymeric coatings (>200 μm thick) to metals, especially steels. The workpieces are preheated and immersed into a bed comprising fluidised polymeric particles, which melt on coming into contact with the pre-heated surfaces, thereby forming a dense polymeric coating. The technique requires significant operator skill. Before coating, it is important to vapour degrease and grit blast the component surfaces, followed by phosphating or chromating. Also see *electrostatic fluidised-bed coating* and *electrostatically sprayed plastic coatings*

fluidised-bed nitriding. See *fluidised-bed thermochemical diffusion methods.*

fluidised-bed nitrocarburising. See *fluidised-bed thermochemical diffusion methods.*

fluidised-bed thermochemical diffusion methods. Fluidised beds can be used to provide rapid thermochemical diffusion treatments; since heat transfer is rapid the time to achieve the required treatment temperature is shorter than for many other thermochemical diffusion methods (except salt bath methods). The most common examples where fluidised bed technology is exploited are nitroarburising, nitriding and carburising; boriding has also received investigation but is not very commonly practised. All these processes utilise a fluidised bed of inert material (often aluminium oxide – corundum) through which appropriate gases (similar to those already described for the equivalent gaseous treatments) are passed. Fluidised-bed operations are intrinsically dusty and stringent working practices are required in order to maintain an acceptable level of workshop cleanliness.

fluxing. A method of removing passive oxide surface layers by the action of aggressive fluxing compounds, e.g., to prepare steel surfaces before galvanising. The activator compound contained in many pack compositions also has a distinct fluxing action.

formability. In the context of surface engineering, the ability of a plastic or metallic coating to adapt to changes in shape without flaking or cracking. A common quality required of coated steel sheet.

fretting. Wear caused through high frequency (kHz), low amplitude (<5 mm) sliding contact.

fretting corrosion. Fretting coupled with the conjoint action of corrosion. For an excellent review refer to the work of Waterhouse cited in the Bibliography.

friction surfacing. A mechanical surface engineering treatment in which a consumable rod is rotated at high speed and applied to a surface under high axial load thereby creating sufficient frictional heating to melt the rod, which becomes layed down as a deposit. The technique can be applied to flat surfaces, plate edges, annular discs and even shafts. The composition of the rods vary but are thought to contain various transition metal cabides. Some are based on AISI 440C martensitic stainless steel (1wt-%C, 17wt-% Cr). The treatment is used to provide wear protection .

frosting. Any surface finishing treatment which produces a surface with a fine matt appearance, e.g., blasting, brushing, barrelling and etching. Sometimes applied to glass for decorative affect.

fused spray coating. See *fusion hard facing alloys.*

fusion coatings. See *fusion hard facing alloys.*

fusion hard facing alloys. Also known as 'self-fluxing overlay coatings'. Applied to an object by firstly thermospraying and secondly fusing with an oxyacetylene torch or an RF induction coil, which 'wets' the coat to the substrate. This produces a coating that is metallurgically bonded to the substrate and is free of microporosity. Hence, it is impervious to corrosive fluids. This two-step method of application is known as the 'spray and fuse process'. There are various alloy types, the most important of which are based on the Ni–Cr–B–Si–C system; depending upon the exact alloy composition, they melt in the range of 980 to 1200°C. The constitution of the Ni–Cr–B–Si–C coatings are complex, but frequently contain relatively large Cr_7C_3 carbide particles (~10–100 μm) in a nickel rich matrix. Some compositions also produce coatings that contain chromium borides. The coatings show excellent resistance to abrasion wear (under light loading) and are reasonably effective in resisting the conjoint action of corrosion and abrasion, e.g., in certain marine applications. Another composition contains additives of coarse (~150μm) tungsten carbide particles which serve to improve abrasion resistance even further.

G

Galfan. The coating produced by Galfanising.

Galfanising. A method of hot dip coating whereby steel, typically in sheet from, is immersed into a bath of molten Zn–5wt%Al alloy held at approximately 450°C. Under optimal processing conditions, a coating free from from massive interfacial intermetallics, is produced. Instead, a fine lamella microstructure, containing η-Zn (a solid solution of iron in Zn) plus θ-$FeAl_3$ and η-Fe_2Al_5 intermetallics, is formed throughout the coating. This method has been developed as an alternative to galvanising. Also see *galvaluming.*

galling. See *seizure.*

galling resistance. See *seizure resistance.*

galvalume. The coating produced by galvaluming.

galvaluming. A method of hot dip coating whereby steel, typically in sheet from, is immersed into a bath of molten 55Al–43.4Zn–1.6Si (wt–%) alloy held at approximately 610°C. The role of the silicon is to retard rapid reaction between the bath and ferrous alloy substrates, which otherwise would produce a coarse distribution of intermetallics. Also see *galfanising.*

galvanic cell. An electrochemical cell having two dissimilar electrical conductors as electrodes.

galvanic contact plating. See *contact plating.*

galvanic corrosion. See *bimetallic corrosion.*

galvanic current. See *bimetallic corrosion.*

galvanic series. A tabulation of metals and alloys in order of their relative potentials in any given environment; but usually sea water. Noble metals like platinum, gold and titanium appear at the top of the listing while base metals like zinc and magnesium appear at the bottom.

galvanising. See *hot dip galvanising*

galvannealing. See *hot dip galvanising*

gaseous aluminising or gaseous calorising. Carried out at temperatures of 950–1050°C in a gaseous medium of aluminium trichloride and hydrogen. Aluminium is reduced at the component surface via the reaction:

$$2AlCl_3 + 3H_2 \longrightarrow 2Al + 6HCl$$

The aluminium is diffused into the metallic components forming various intermetallic compounds. Forced circulation of the gases improves uniformity of the treatment throughout a production charge. The process plant required is practically the same as that used for CVD. Also see *aluminising.*

gaseous austenitic nitrocarburising. Conducted in sealed quench furnaces using a mixture of endothermic and ammonia gases and generally carried out at ~ 700°C. Typical transformed austenite case thicknesses are in the range of 50–200µm with a Vickers hardness ~750–900 kg/mm^2.

Case strength is developed by quenching (to obtain a nitrogen–carbon martensite) after nitrocarburising which is further enhanced by ageing to decompose retained austenite to nitrogen bainite. Apart from improving contact load resistance, fatigue strength can be improved by as much as 100%. Proprietry gaseous austenitic nitrocarburising treatments are marketed as *Nitrotec C, Alpha Plus* and *Beta.*

gaseous boriding.

'Boriding carried out in a gaseous medium' – IFHT DEFINITION.

Boriding carried out in a gaseous medium containing, typically, boron trichloride (BCl_3) mixed with hydrogen and (sometimes) nitrogen. The process maybe carried out at atmospheric or sub-atmospheric pressures. Temperatures and times are the same as those cited under *boriding*. Also see the review: P . A .Dearnley and T. Bell, *Surface Engineering*, 1985, **1**, (3), 203-217.

gaseous boroaluminising. A multicomponent boriding technique resulting in the *simultaneous* diffusion of boron and aluminium into a steel surface.The gaseous medium is formed by passing HCl gas through a pack containing ferroboron, aluminium and silicon carbide. The process is conducted at temperatures ~950-1100°C for up to 3.5 hours. It is less often practised than the equivalent pack method and does not provide the optimal distribution of boron and aluminium, which only *sequential* diffusion can provide. However, there is no reason why this process could not be appropriately developed to enable sequential diffusion. Also see *multicomponent boriding.*

gaseous calorising. See *gaseous aluminising.*

gaseous carbonitriding.

'Carbonitriding carried out in a gaseous medium' – IFHT DEFINITION.

Essentially a modified form of gaseous carburising in which ammonia is introduced into the carburising atmosphere. Only sufficient ammonia is added to obtain the required improvement in case hardenability. However, slightly higher nitrogen levels can enable an overall improvement in temper resistance. Also see *carbonitriding* for general comments on treatment temperature and case depth.

gaseous carburising

'Carburising carried out in a gaseous medium' – IFHT DEFINITION.

Probably the most popular industrial carburising method, usually carried out at temperatures ≈925-950°C. Batch or continuous furnace designs are the most commonly used. Compared to plasma and vacuum carburising, this process lies closer to thermodynamic equilibrium. Nonetheless, although possible, thermodynamic equilibrium rarely prevails in industrial gaseous carburising systems, and the use of thermodynamic theory has limited usefulness for process control purposes.

Endothermic gas (endogas) is the principal carbon source used in gas carburising. It is a mixture of carbon monoxide(15-20%), hydrogen (35-45%) and nitrogen (35-45%) with smaller amounts of carbon dioxide(0-1%), methane (0.5-1.5%) and water vapour; it is generated on site by the combustion of air-propane mixtures and has a carbon potential ranging between 0.35 and 0.50wt-%C. Before entering the furnace an addition of methane (CH_4), propane (C_3H_8), butane (C_4H_{10}), methanol (CH_3OH) or ethanol (C_2H_5OH) is added which further increases the carbon potential to a level required for carburising, usually around 0.8 wt-%. The endogas is said to act as a "carrier gas" for the hydrocarbon additive. Liquid additives like methanol, or more complex proprietry mixtures (like glycols, ketones and benzene) are admitted as droplets which fall onto a heated plate, inside the furnace, and are vapourised and carried into the furnace by the endogas. The latter procedure is sometimes called the drip feed method.

Since many gases exist in the furnace atmosphere there are several reactions that take place. Indeed, the precise reactions are still a subject of some contention. The following, however, are believed to be the main ones involved in the mass transfer of carbon to the steel surface.

$$2CO \longrightarrow C_{(Fe)} + CO_2 \qquad (1)$$

$$CO + H_2 \longrightarrow C_{(Fe)} + H_2O \qquad (2)$$

$$CH_4 \longrightarrow C_{(Fe)} + 2H_2 \qquad (3)$$

$$CO \longrightarrow C_{(Fe)} + 1/2\ O_2 \qquad (4)$$

Although all the above reactions have been shown moving in the direction of carburising, it should be appreciated that they are all of the reversible type. Reaction 1 is sometimes known as the Boudouard reaction, while reaction 2 is called the water gas reaction. Since the water content of the furnace gases can be determined by using the dew point method, it is possible to estimate the carbon potential of the furnace. Carbon potential is then "fine tuned" by varying the quantity of hydrocarbon additive. Greater precision in the measurement of carbon potential can be obtained by using an *infra-red gas analyser* which is able to directly measure the quantity of CO and CO_2. In fact from equation 1 it can be appreciated that the carbon potential is proportional to $(pCO)^2/pCO_2$. Yet another method of asssessing carbon potential is to use a zirconium oxygen sensor (sometimes called an oxygen probe) which detects oxygen generated via reaction 4. Here, the carbon potential is proportional to $(pCO)/p(O_2)^{1/2}$. Also see *zirconia oxygen sensor.*

gaseous chromising

'Chromising carried out in a gaseous medium' – IFHT DEFINITION.

In gaseous chromising a chromium halide, usually $CrCl_2$, acts as the chromium source. It is reduced in the gaseous state by hydrogen:

$$CrCl_2 + H_2 \longrightarrow Cr + 2HCl$$

and also at the steel substrate surface (by an exchange reaction):

$$CrCl_2 + Fe \longrightarrow Cr + FeCl_2$$

$CrCl_2$ is often generated by passing HCl over chromium powder heated at 850-1000°C, which is connected directly to the chromising vessel. Gaseous chromising is carried out at 850 and 1050°C for durations up to 12 hours. Chromium potential can be more precisely controlled compared with the pack method. It is important to control the rate of $CrCl_2$ reduction such that it does not exceed the rate of chromium solution by the substrates, otherwise a surface layer of uncombined chromium will be produced. The rate of chromium solution decreases as the carbon content of a given steel substrate increases.

gaseous nitriding

'Nitriding carried out in a gaseous medium' – IFHT DEFINITION.

In conventional gaseous nitriding, pure anhydrous ammonia acts as the nitriding medium and a single treatment temperature, typically ≈500°C, is usually employed. The ammonia is catalytically dissociated as it passes over the prior hardened and tempered low alloy steel charge, according to the reaction:

$$NH_3 \longrightarrow 3/2H_2 + N_{(Fe)}$$

In industrial furnaces, additional dissociation takes place on the chamber walls and associated metallic parts. Nitrogen potential is controlled by controlling the flow rate of ammonia over the charge; this controls the amount of dissociation. A level of 15 to 30% dissociation is usually maintained in the exhaust gas. Nowadays dissociation is best measured using an infra-red gas analyser, although, absorbtion pipettes are still in use. At the beginning and end of the treatments, during heating and cooling, it is usual to purge the system with nitrogen.

It has been experienced that the regulation of nitrogen potential through the flow rate of ammonia alone is too crude to enable precise control of nitrogen potential, e.g., as is required in bright nitriding. A far better option is to mix hydrogen with ammonia and to control nitrogen potential by controlling the ratio of these two gases. This principle was first demonstrated in laboratory experiments by Lehrer (circa 1930) but it was not until the 1970s that the same principle was implemented on an industrial scale. However, a similar effect was exploited by the Floe Process (also called double-stage nitriding), developed in the USA, circa 1948. This proceeds like conventional gas nitriding for the first few hours, after which dissociated ammonia is added to the anhydrous ammonia to achieve exhaust gas dissociation levels of 75 to 80%, i.e., the second part of the nitriding cycle is conducted at considerably reduced nitrogen potential. A slightly higher nitriding temperature may also be used during the second stage (≈550-565°C) compared to the first (≈500-525°C). This process is still practised in the USA. Despite such refinements, however, a surprising amount of industrial gaseous nitriding is reliant on anhydrous ammonia alone. Also see *nitrogen potential* and *zirconia oxygen sensor.*

gaseous nitrocarburising

'Nitrocarburising carried out in a gaseous medium'
– IFHT DEFINITION.

gaseous sherardising. Sherardising carried out in a gaseous medium containing zinc halides at temperatures ~300–500°C. In view of the complexity of the equipment required, gas sherardising is not economically justified, since the products requiring sheradising are of very low unit cost. Also see general comments under *sheradising*.

gaseous siliconising

'Siliconising carried out in a gaseous medium'
– IFHT DEFINITION.

Carried out by the reduction of gaseous $SiCl_4$ by H_2, or the pyrolysis of silane (SiH_4). The process is conducted at temperatures ~800-1200°C for 6-7 hours. Also see *siliconising*.

gas plating. See *chemical vapour deposition (CVD).*

GDOES. See *glow discharge optical emission spectroscopy.*

gilding. (i) A traditional term for flash gold (24 carat) plating of fancy goods, trinkets and giftware sold at the cheaper end of the domestic market. See *gold plating*. (ii) The mechanical application of gold leaf (0.076 to 0.127 μm thick) to wooden or base metal objects.

glancing angle X-ray diffraction. A variant on X-ray diffraction used for obtaining crystal spacing (*d*) data from the near surface. It differs from the usual X-ray methods in that the incident X-ray beam is made to intercept the sample surface at some small glancing angle (γ). There are two methods: (i) the Read camera (ii) the Seeman–Bohlin diffractometer. The former is a film method (diagram) while the latter is a specially configured X-ray goniometer that uses a proportional counter to measure diffraction intensity; the output is a plot of intensity versus 4θ, where θ is the Bragg angle. A glancing angle of 6.4° increases the X-ray path length, in the near surface, by about nine times. Very thin (~1 μm) coatings can be interrogated by this method.

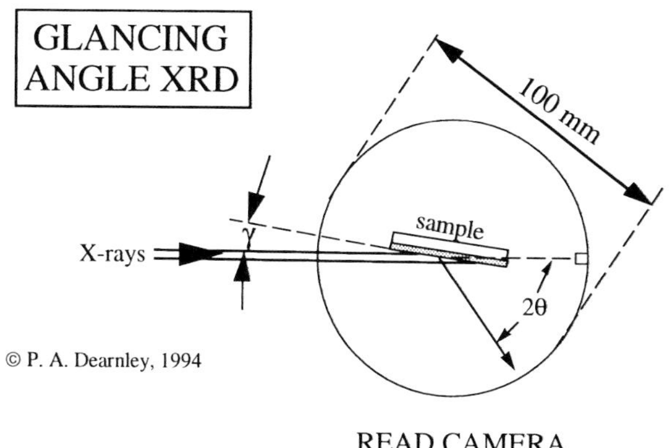

READ CAMERA

glass structure. See *amorphous structure.*

glow discharge boriding. See *plasma boriding.*

glow discharge carburising. See *plasma carburising.*

glow discharge heat treatment. See *plasma heat treatment.*

glow discharge nitriding. See *plasma nitriding.*

glow discharge optical emission spectroscopy (GDOES). A relatively recent (circa 1979) optical emission spectroscopy technique that has the capability to analyse a large number of elements. The design of the sample chamber is based on the Grimm lamp (diagram). The sample is made cathodic with respect to an anular anode (~ 8mm in diameter) through which a low pressure flow of argon is passed. A current intensive glow discharge plasma is subsequently created at a pressure ~ 1.5 torr and material is removed from the sample face by high rate sputtering. Whilst in-flight the excited sputtered sample atoms emit photons of characteristic wave-length (due to electronic transitions); see diagram. The light is then focused and passed through a collimating slit into a spectrometer where it strikes a concave holographic grating, splitting the light into its component wave-lengths. The spectral detection range is ~110 to 1000 nm. The system must be calibrated against primary standards, but can routinely analyse all the major elements including, hydrogen, nitrogen, carbon, oxygen and boron. It also has the capability to achieve rapid depth profiling, making the technique of special value for the analysis of ceramic coatings and diffusion zones. For example, it is possible to analyse 10 elements to a depth of 50 μm in less than 30 minutes. However, some experiences show that depth resolution is not better than 0.5μm.

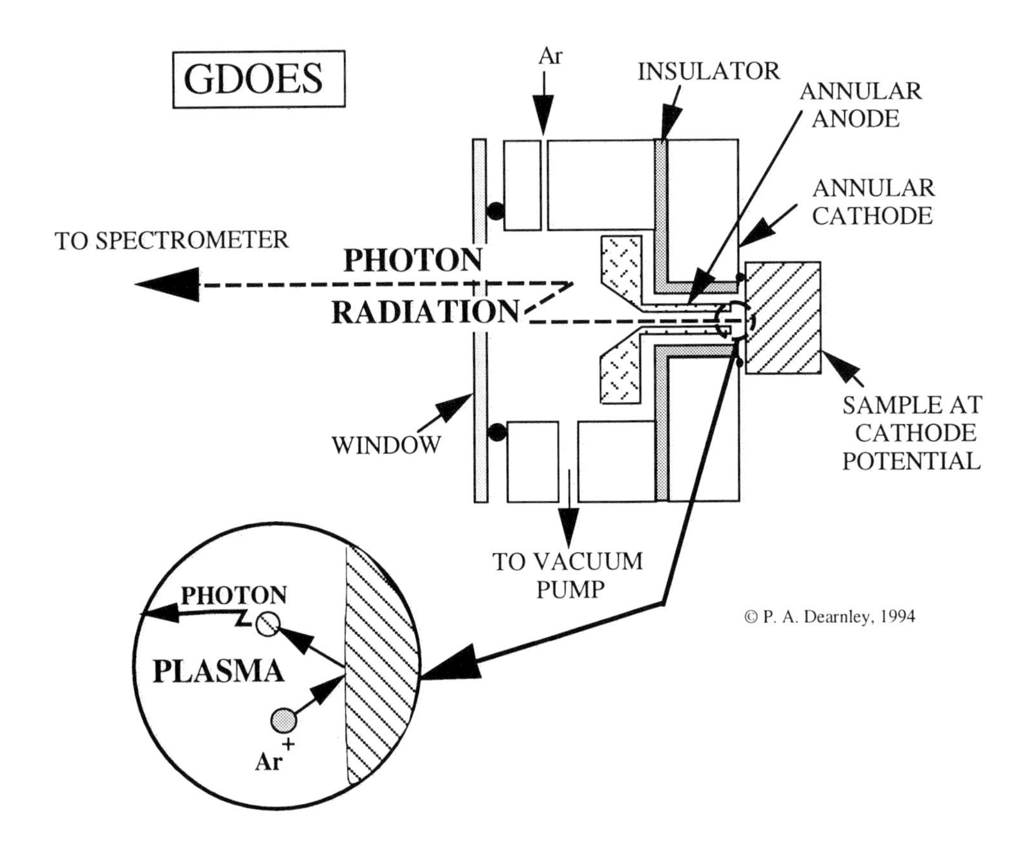

glow discharge plasma. Usually referring to a DC (direct current) glow discharge plasma. When a potential of a few hundred volts is applied across two electrodes held in a low pressure (~10^{-2} to 10 torr) gas, partial ionisation of the gas occurs, causing ions to move towards the cathode and electrons towards the anode. In this way electrical energy is passed through a gas. Bernhard Berghaus was the first scientist to systematically explore the industrial surface engineering potential of glow discharge plasmas and filed a number of seminal patents on sputter deposition and plasma nitriding. The essential features of a DC glow discharge plasma are shown in the graphics diagram. In industrial systems the chamber wall serves as the anode which is held at ground potential. Plasma based surface engineering processes like plasma nitriding, plasma carburising and magnetron sputtering utilise plasma power densities up to 15 W/cm^2. Such techniques require the use of power supplies equipped with arc suppresion control. See *arc suppression*. When insulating surfaces require treatment radio frequency glow discharges are used. See *radio frequency glow discharge*. See also ***colour section***, p. A.

GLOW DISCHARGE PLASMA

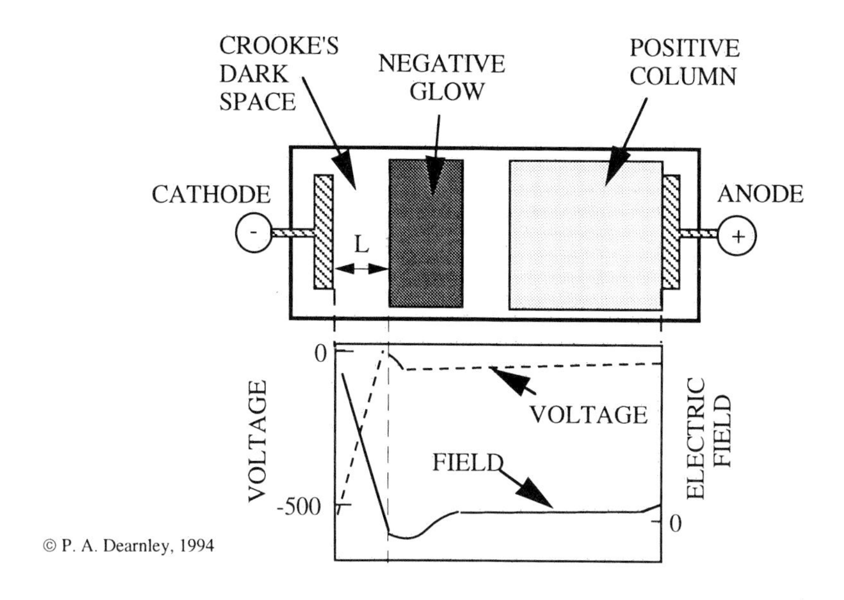

glow discharge PVD. See *plasma assisted PVD (physical vapour deposition).*

glow discharge siliconising. See *plasma siliconising.*

glow discharge titanising. See *plasma assisted CVD.*

gold plating. Base metals, like zinc and brass castings, are given a prior bright nickel electroplate deposit. Gold plating (gilding) is costly and hence electroplating is carried out for very short times (typically ~5 to 15 seconds). Hence, the term flash gilding. For filigree work (delicate ornamental objects made from wire) thicker coatings are needed and electroplating is therefore carried out for 20 to 30 seconds.

gradated coating. Any coating whose constitution varies continuously between successive layers, usually enabling a progressive blending of properties between that of the substrate and the outermost layers of the coating. Coatings of this type can be produced by thermal spraying or plasma assisted PVD methods.

graphitic corrosion. Corrosion of grey cast iron, characterised by preferential solution of the matrix, leaving behind unreacted graphite.

green rot. The oxidation or carburising of certain nickel alloys at ~1000°C that results in the formation of a green residue.

grit blasting. A process for removing rust, paint or unwanted surface deposits (e.g., flash) from components. A high velocity air stream is used to propel angular shaped alumina or silicon carbide particles onto component surfaces at high speed. The flow is directed through a flexible hose and nozzle, enabling manipulation of the flow direction. Grit blasting is a common preparation stage prior to thermal spray coating.

H

hard chromium plating. See *chromium plating*

hard coating. A term usually referring to transition metal carbides and nitrides deposited by CVD or plasma assisted PVD methods, e.g., TiC and TiN, which have Vickers hardness values in excesss of 2000 kg/mm^2. The term hard coating tends to be invoked when these coatings are being applied in order to improve wear resistance of various items, e.g., metal cutting tools or bio-medical prosthetic implants.

hardening

'Any treatment designed to render an object significantly harder'
– IFHT DEFINITION.
Also see *surface hardening.*

hardfacing. See *weld hardfacing.*

hardness. Qualitatively, a measure of the resistance of a surface to penetration by an indenter. Quantitatively, a measure of yield strength. For example, Vickers hardness *Hv* is related to yield strength (σ_y) by the approximate relationships:

$$Hv \approx 3\ \sigma_y \text{ (for metals and alloys)}$$
$$Hv \approx 4\ \sigma_y \text{ (for ceramics)}$$

hardness distribution. See *hardness profile.*

hardness measurement. Various methods of hardness determination exist. These can be broadly grouped into: (i) static and; (ii) dynamic hardness methods. In surface engineering only static methods are used. These comprise Vickers, Knoop and Berkovich diamond indentation methods. Rockwell hardness on scales A, B or C is used in accordance with the type of material; most popular in the United States and unsuited to microhardness determination. See *Vickers hardness, Knoop hardness,* and *Berkovich indentation hardness.*

hardness profile

> 'Hardness as a function of distance from a fixed reference point (usually from a surface)' – IFHT DEFINITION.

Also sometimes termed hardness distribution. In surface engineering it specifically refers to *microhardness* as a function of depth below the surface. The shape and magnitude of such curves are often a signature of a given type of treatment. Reverse 'S' shaped curves are a characteristic feature of nitrided or carburised low alloy steels (see diagram) while single step-like curves are typical of hard coated ferrous or non-ferrous alloys or nitrided high alloy steels. Hardness profiles are most easily determined using Vickers or Knoop hardness indentation methods. Also see *Vickers hardness, Knoop hardness, Berkovich indentation hardness* and *nanoindentation hardness.*

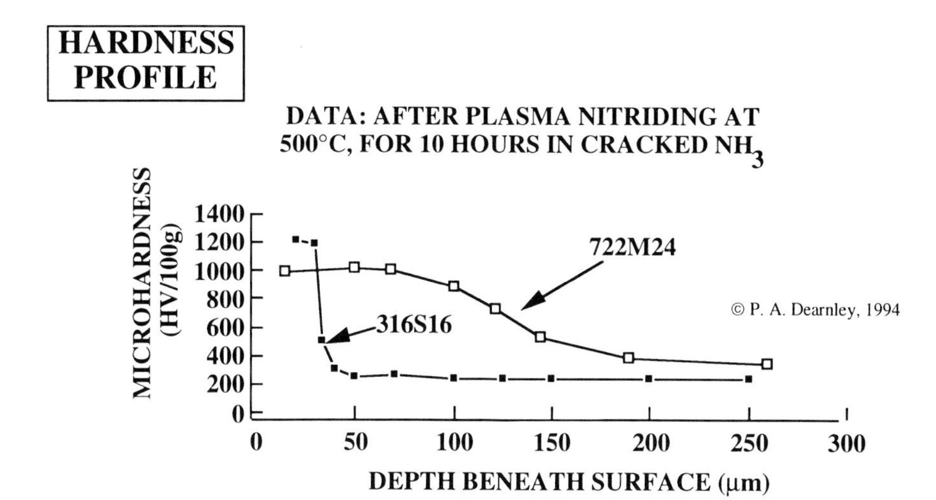

Note: 316S16 is an 18-8 austenitic stainless steel; 722M24 is a low alloy nitriding steel.

hard nickel plating. Nickel electroplating carried out in a modified Watts bath called a 'hard Watts bath'. The electrolyte contains organic or ammonium ion additions that result in modification of the coating structure, causing an increase in hardness (~350–500 kg/mm^2), relative to standard nickel electrodeposits (~150–200 kg/mm^2), but at the expense of ductility and toughness, e.g., the strain to fracture of standard nickel electrodeposits are ~ 0.2–0.3, while values for hard nickel are ~0.05–0.08. Hard nickel plating is used in engineering applications where wear *and* corrosion resistance is required, over and above that provided by standard nickel electrodeposits.

hard plating. Jargon for hard chromium or hard nickel plating, but usually the former.

hard surfacing. See *weld hardfacing*.

heat-affected zone (HAZ). In surface engineering this refers to the HAZ beneath a power beam alloyed or clad surface, i.e., the zone, adjacent the formerly liquid region, which remained in the solid state for the entirety of the treatment, but whose microstructure has been changed from that of the core, as a result of rapid heating and cooling.

heat stability. See *thermal stability.*

heat treatment. The manipulation of bulk and surface properties of a material by the concise application of heating and cooling cycles, in an appropriate atmosphere. For steels, common bulk heat treatments include, annealing, hardening & tempering, normalising, stress relieving and sub-critical annealing. Some common 'surface heat treatments', include carburising, nitriding, nitrocarburising and boriding.

Hertzian cracks. Cone cracks formed on the surface of a brittle elastic solid (e.g., ceramics and glass) resulting from the high contact stresses exerted during point contact loading by a spherical indenter (or similar). Note: in sliding contacts ring shaped cracks are replaced by a series of overlapping arc-shaped cracks. This effect can also be seen following scratch testing.

Hertzian failure. A situation whereby a high yield strength and/or high modulus coating fails through plastic flow of a low yield strength substrate during intensive point contact loading (diagram and micrograph). Sometimes referred to as the 'thin ice effect'. Also see *Hertzian stresses.*

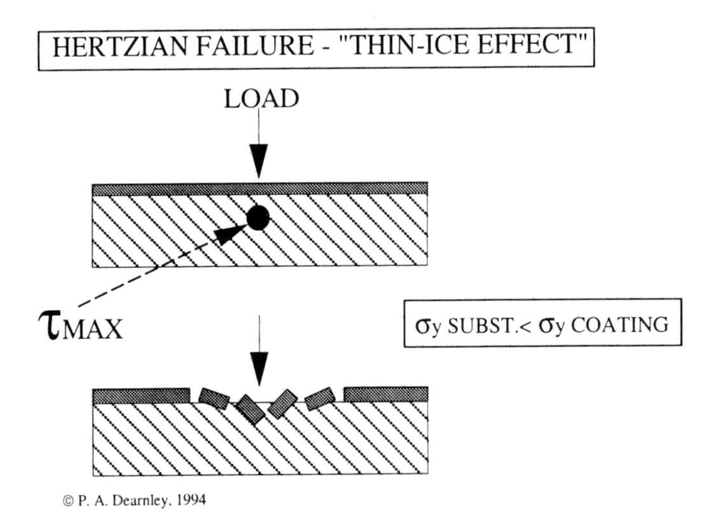

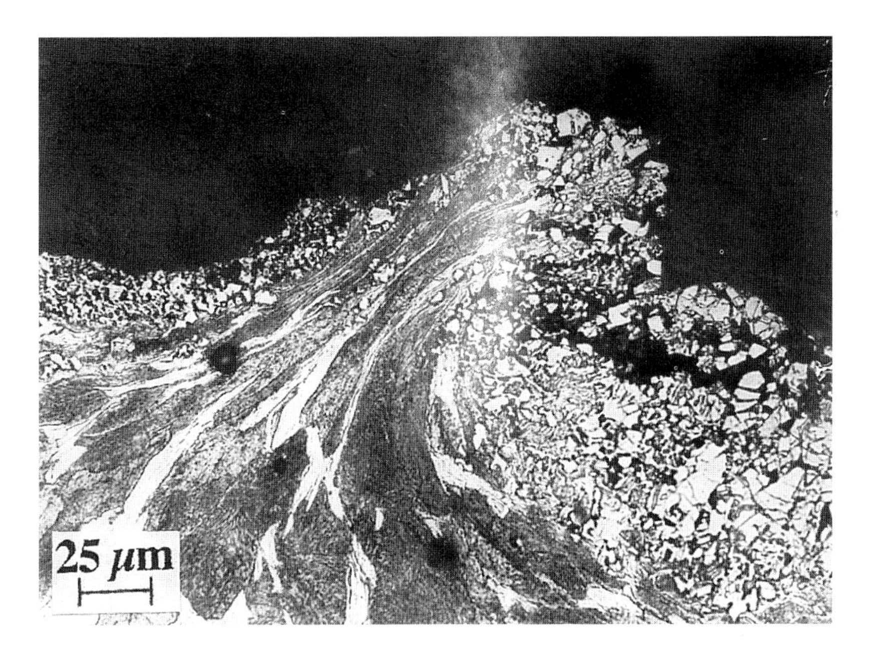

Hertzian failure. *The Achilles heel of hard coatings on 'soft' substrates. The micrograph is a section prepared through a boronised 0.4% C steel test wheel after undergoing an Amsler wear test with an applied normal load of 1500 N under rolling contact with 10% slip. Massive subsurface plastic flow of the steel substrate has occurred and the hard (≈ 1700 kg/mm²) exterior iron boride layer has fragmented through brittle fracture. Light optical micrograph (etched in 2% nital). From P. A. Dearnley, unpublished research, 1987.*

Hertzian stresses. The Hertzian elastic contact stresses developed, for example, when an indenter contacts a planar surface. The shape of the elastic stress field and the position of the maximum resolved shear stress depends upon indenter geometry, while the magnitude of the compressive and shear stresses (for a given contact force) is dependent upon the elastic moduli and Poissons ratios of the indenter and planar materials. For a spherical indenter, indenting a planar surface, the radius of circular contact (a) is given by:

$$a = (3FR/4E^*)^{1/3}$$

where:

F = applied load (N)
R = radius of the indenter (m)
$E^* = [(1-\upsilon_1^2/E_1) + (1-\upsilon_2^2/E_2)]^{-1}$
E_1 = indenter Young's modulus (GPa); υ_1 = Poissons ratio of indenter
E_2 = Young's modulus of the semi-infinite surface (GPa);
υ_2 = Poisson's ratio of semi-infinite surface

The maximum contact pressure P_m (MPa) at the contact interface is given by:

$$P_m = (3F/2\pi a^2) = (6FE^{*2}/\pi^3 R^2)^{1/3}$$

The shear stress τ is zero at the contact interface but achieves its maximum value along the indenter centre line at a position that is exactly 0.48a beneath the planar surface. The magnitude of τ_{max} (MPa) is simply related to P_m:

$$\tau_{max} = 0.31\ P_m$$

For a cylinder contacting a planar surface the Hertzian equations are modified, but it is worth noting that $\tau_{max} = 0.30\ P_m$ (very close to the relationship for a spherical indenter), whereas, the position of the maximum shear stress, is exactly 0.78a beneath the planar surface. Hence, when considering tribological situations where third body particles are present, the particle (or indenter) shape has a very strong influence on the *position* of the maximum shear stress. This has implications for coating design specifications. Coating 'X' maybe of adequate thickness for avoiding sub-surface yielding by spherical particles, but maybe inadequate for avoiding sub-surface yielding by cylindrical or rod-shaped particles. For a more rigorous mathematical analysis the reader is referred to K. L. Johnson's *Contact Mechanics,* Cambridge University Press, 1987.

heteroepitaxy. See *epitaxy.*

hexavalent chromium ion. The most common valency state for ionic chromium involved in the electrodeposition of chromium. Also see *trivalent chromium plating.*

high energy beams. A qualitative term referring to high power density heat sources, like lasers and electron beams.

high-frequency hardening. See *induction hardening.*

high-power laser. A qualitative term, given that the power capability of lasers is being continuously upgraded. In general, however, those lasers deployed for surface engineering, regarded as high-power, are capable of delivering power densities in the range of 10^3 to 10^9 W/cm^2. These should be distinguished from low power lasers used for metrology and surveying purposes. Also see *laser.*

high-temperature carburising. Carburising carried out at temperatures above 950°C, (usually 1000–1100°C). This is typical for vacuum and plasma carburising methods.

high-temperature galvanising. Hot dip galvanising performed at a higher than normal temperature (e.g. at 550°C). The method is more tolerant to variations in dipping time than conventional hot dip galvanising and necessitates the use of a ceramic vessel.

high-temperature stability. See *thermal stability.*

high velocity air fuel (HVAF) spraying. A similar concept to high velocity oxygen fuel (HVOF) spraying except this particular design of torch has the economic advantage of running on a mixture of air and kerosene (paraffin). Particle velocities are claimed to be equivalent to those obtained with HVOF. This technology has been devised by the Browning Corporation (USA). A variety of coatings can be deposited, but WC-Co is one of the more common. Also see *detonation gun spraying* and *high velocity oxygen fuel (HVOF) spraying*.

high velocity oxygen fuel (HVOF) spraying. Originally developed by the Browning Corporation (USA) this process pre-mixes oxygen and a fuel gas like acetylene and passes them into a flame nozzle where they are ignited to produce a ring of 'flame jets'. With the help of a carrier gas, coating powder is injected along the the central axis of the combustion flame. Very high particle velocities are achieved. There is some contention regarding the precise particle velocities but these are generally agreed to be several times those achievable with plasma spraying. One claim is that they reach ~1000 m/s, which should be compared with ~150 m/s for plasma spraying. It is claimed that the coatings produced are of comparable density, if not superior, to vacuum plasma sprayed coatings. The manufacturers of HVOF and HVAF systems also state that the powder particles receive significant heating on impacting the component surface. There is contention regarding the relative proportions of heat imparted to the powder particles during impact and that imparted by the torch 'flame'. A variety of coatings can be deposited, but WC-Co is one of the more common. One commercial torch of this type goes by the name of 'Jet-Kote'. Also see *detonation gun spraying* and *high velocity air fuel (HVAF) spraying*.

holiday test. A quality control test used for determining surface connected porosity or microdefects in paint or plastic film (if applied to metallic objects). The substrate is connected to one side of a DC power supply, while a high voltage (low power) metal probe is passed over the treated surface. Any micro-defects in the coating results in the closure of the electrical circuit, and the triggering of an audible or visual signal.

hollow cathode. A current intensive effect observed when two cathode plates, surrounded by a glow discharge plasma, come into close proximity, such that their respecive dark spaces or glow seams overlap. The situation may also arise inside a hole within a negatively charged component, as for example during plasma nitriding. The magnitude of the discharge current is inversely proportional to the cathode plate separation (see diagram below, based on data from N. A. G. Ahmed and D. G. Teer, *Thin Solid Films*, 1981, **80**, 49-54) or hole diameter. At some critical spacing, current densities can be sufficient to melt many metals in a matter of a few seconds. With experience, however, such catastrophies can be avoided. A useful exploitation of the hollow cathode effect is the hollow cathode source evaporator. See *hollow cathode source evaporative PVD*.

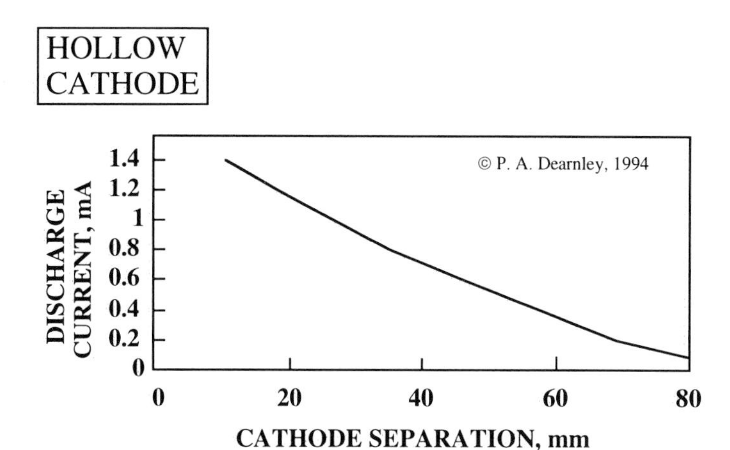

hollow cathode source evaporative PVD. An evaporation source PVD process in which a cyindrical cathode (fabricated from a refractory metal) is used to generate a hollow cathode which emits an intense beam of electrons. The beam maybe collimated or diffuse and can be steered and focused onto the evaporant with an external magnetic field. Such devices have proven popular in a number of Japanese plasma assisted PVD units. They are relatively less complicated than electron beam evaporators and do not require independent evacuation. Also see *ion plating.*

homoepitaxy. See *epitaxy.*

hot dip aluminising. Immersion of pickled and pre-fluxed steel plate into molten Al-Si alloy (2 to 10wt-% Si) at 620-710°C to produce a protective surface coating. The silicon serves to retard the growth rate of the interfacial intermetallic compound, η-Fe_2Al_5, which has a characteristic "saw-tooth" morphology, resembling Fe_2B layers fromed on borided plain carbon steel. The outer surface of the coating is rich in pure aluminium. Hot dip alloys containing 0-2wt-% Si are preferred for corrosion resisting applications while 5-10 wt-%Si is specified for oxidation resisting applications. After hot dipping, an annealing or diffusion treatment may be carried out to obviate spalling and to further improve high-temperature oxidation or corrosion resistance. Coatings are generally ~25 to 75 μm thick. Sometimes hot dip aluminising is used as a pre-treatment for *vitreous enamel coating.*

hot dip galvanising. Immersion of steel plate (usually) into molten zinc or Zn-Al alloy (0.1 to 0.3wt-% Al) to produce a protective surface coating some 80 to 125 μm thick. The bath temperature is normally 430-460°C. Pre-treatment of the steel may include annealing in a reducing atmosphere, acid-pickling and fluxing with zinc chloride. After hot dip galvanising, an annealing or diffusion treatment (sometimes called galvannealing) may be carried out to substantially improve outdoor corrosion resistance, spot weldability and paintability. Galvanised layers have a complex constitution but often comprise a mixture of γ–Fe_5Zn_{21},

δ–$FeZn_7$ and ξ–$FeZn_{13}$ and η-Zn (a solid solution of iron in Zn). Generally, the γ-phase comprises <5% the thickness of the total layer, the remainder being dominated by δ, ξ and η. Some workers give γ–Fe_5Zn_{21} as Γ_1-Fe_5Zn_{21} and suggest the phases Γ-Fe_3Zn_{10} and δ_1–$FeZn_{10}$ are also important. For further details see J. Mackowiak and N. R. Short, *Int. Met. Rev.*, 1979, **24,** 1-19.

hot dip tinning. Hot dip tin has a corrosion resistance similar to that of zinc and, because of its low toxicity, is used for the corrosion protection of steel used in the food processing industry. Processing steps are similar to galvanising; the metal sheet, for example, is prepared by pickling and fluxing followed by immersion in molten tin for a short time. Steels, copper and copper alloys can be hot dipped in this way. Deposits are ~12-50 µm thick.

hot dip zinc-aluminium galvanising. See *galfanising* and *galvaluming.*

hot filament CVD. A CVD process in which a hot (typically 2000°C) tungsten filament, is placed near the surface of an object, so as to create a localised plasma. This approach has been used for the deposition of microcrystalline diamond coatings, which uses an appropriate hydrocarbon gas, but is presently less popular than *microwave plasma CVD.* Also see *diamond coatings* and *diamond-like coatings.*

HVAF. See *high velocity air fuel spraying*

HVOF. See *high velocity oxygen fuel spraying*

hydrodynamic lubrication. A lubrication situation whereby two sliding surfaces are separated by a relatively thick film of lubricant and the normal load is supported by the pressure within the film which is created hydrodynamically. Both sliding surfaces must have conforming geometries to enable the establishment of hydrodynamic lubrication. Contrast with *elastohydrodynamic lubrication (EHL).*

hydrogen electrode. An electrode at which the equilibrium reaction H^+ (aq.) + e —> 1/2 H_2 is set up. The potential of the standard hydrogen electrode = 0.00 volts.

hydrogen embrittlement. A problem encountered in chromium or nickel plated steel objects, whereby significant quantities of hydrogen, evolved during electroplating or prior alkaline or acid cleaning/pickling, becomes dissolved in the steel and impairs toughness and ductility. Such hydrogen can be removed, or redistributed, by 'baking' to ~200°C. However, if *fatigue* strength is of critical importance, baking must be carried out above 450°C, or not at all, since baking at 200°C causes a major reduction in fatigue strength, even compared with 'unbaked' material.

hypersonic powder spraying. See *HVAF* and *HVOF.*

I

Ihrigising. See *siliconising.*

immersion cleaning. Cleaning performed by the chemical action of a fluid into which the object to be cleaned is dipped.

immersion gilding. Small articles made of brass or copper can be given a flash coating of gold by a simple immersion method. The gold is deposited by a displacement reaction. Immersion gilded items include domestic fashionware, printed circuit boards and electrical bridging pieces.

immersion plating. See *autocatalytic plating.*

impact surface treatment. A mechanical treatment in which the surface is hardened by repetitive cold working, e.g., by bombarding with small hard particles (peening) or by hammering.

implantation. See *ion implantation.*

implantation dose. Quantity of ions implanted into a surface, typically residing in the range of 10^{17}-10^{18} ions/cm^2. See *ion implantation.*

implantation induced amorphisation. The onset of surface amorphisation observed in some systems when the ion implantation dose exceeds a critical value.

implantation induced defects. Surface defects resulting from improper ion implantation. These include: (i) excessive surface roughening caused by sputtering and; (ii) blistering and swelling, due to coalescence of implanted gases.

implant concentration profile. The variation in implantation concentration as a function of depth beneath the surface. Usually, this takes the form of a skewed Gausian curve with the maxima some 100nm or so beneath the surface.

implanted layer. The near surface zone whose composition has been modified by ion implantation with an element which differs from the substrate matrix.

impressed current cathodic protection. See *cathodic protection.*

impulse plasma. See *pulse plasma* .

indentation adhesion test. See *adhesive (or adhesion) strength tests.*

indium plating. Indium has a very low melting point (156°C) and consequently has a very greasy or lubricous 'feel'. Accordingly, it has found application as a surface coating for high performance bearings, e.g., for use at the big-ends and main-ends of petrol and diesel engines. Such bearings comprise a very low carbon steel substrate coated with lead, leaded-bronze, copper-lead, bronze-lead or cadmium. The indium is electrodeposited from a proprietry 'acid indium' solution for a relatively short time; the resulting layer is only 3-4% the thickness of the coating beneath. The indium is subsequently diffused into the sub-layer beneath, by heating to 170°C for two hours. Alternatively, indium plating may serve as a bond coating, e.g., see *bearing shells.*

induction hardening

> 'Quench-hardening treatment involving induction heating, generally used for surface hardening – IFHT DEFINITION.

A rapid method of surface hardening applied to steels of medium carbon content ≈0.3 to 0.5 wt-%. The steel should be at least in the normalised condition, but preferably hardened and tempered, providing sufficient bulk strength for heavy duty applications. Fully annealed steels are unsatisfactory for most purposes. A high frequency induction coil (0.5 to 500 k.Hz) is used to induce eddy currents into the steel surface which causes rapid heating. There is much craft, as well as science, in the design of a given induction coil for a particular application, and most heat treatment shops have a range of coils for various purposes. Immediately following austenitisation of the surface the component is allowed to self quench (by heat transfer into the relatively cooler core) or maybe quenched by appropriately positioned water jets. After hardening, tempering at ~150 to 200°C is carried out to avoid excessive brittleness. This is especially important if the parts are to be subsequently ground (to obviate cracking).

The power input of induction coils lies in the range of 0.1 to 2.0 kW/cm^2. Choice of frequency depends upon the depth of treatment required per second of interaction time. Simplified relationships exist to enable calculation of the approximate depth of the heat affected zone (d_h) for steel up to the hardening temperature. For example, at 800°C; $d_h \approx 100/\sqrt{f}$, where f is the induction frequency in Hz. Hence, high frequencies ~ 250 k.Hz are used where shallow treatments ~ 0.2 mm are required, while low frequencies ~ 4 k.Hz are used where deeper treatments ~ 1.6 mm are required. Induction coil diameter and the number of turns in a particular coil, are also of importance. Also see *flame hardening*.

induction tempering. Surface tempering using induction heating. The tempering time in this process is extremely short and localised. For large surface areas, furnace tempering is mandatory.

infra-red gas analyser. Gas analyser of the absorbtion kind, factory set to determine ammonia or CO/CO_2 levels in gaseous nitriding and carburising furnaces respectively. It works on the principle that when infra-red radiation (λ = 2 to 11 μm) passes through a gas certain gases absorb the radiation in accordance with the amount of gas present. Elemental gases do not absorb infra-red and hence cannot be detected by this technique; a mass spectrometer would have to be used in such cases.

infra-red reflectance. The ability of a surface to reflect infrared radiation. See *reflectance.*

inhibitor. Any substance added to an environment in relatively small quantities which serves to reduce the corrosion rate of objects placed in it.

interchange CVD reaction. See *displacement CVD reaction.*

intermediate coating. See sub-layer.

internal oxidation. Oxidation involving precipitation of dispersed oxides within a metal sub-surface, while oxygen is supplied from the surface via diffusion. For steels, internal oxidation is analogous to nitriding.

internal stresses. See *residual stresses.*

interrupted electroplating. See *pulse plating.*

ion assisted surface treatments. Any method that deploys an ion beam or plasma that (directly or indirectly) achieves a change in surface properties.

ion beam. Any beam of ions generated by any method. Energies may vary widely depending upon the application, from ~100 ev to ~100 kev. For example see *ion implantation.*

ion beam alloying. See *ion implantation* and *ion beam mixing.*

ion beam assisted deposition. A hybridised process, whereby a plasma assisted or non-plasma PVD process is supplemented by an ion beam which is used to bombard the substrate surface before and/or during deposition. Sometimes used in conjunction with balanced magnetron sputter deposition techniques to improve coating adhesion. This procedure is now obviated, to some extent, by the invention of the unbalanced magnetron sputtering source. Also see *ion beam sputter deposition.*

ion beam energy. See *ion beam.*

ion beam enhanced deposition. See *ion beam assisted deposition.*

ion beam irradiation. The exposure of a solid surface to an ion beam.

ion beam mixing. (i) A variant on ion implantation, whereby a thin (~0.1μm) preplaced coating is irridiated by high energy ions (usually nitrogen), causing the coating material to enter the substrate; (ii) ion implantation with more than one element.

ion beam source. Any device for ion beam generation. Many designs exist. These include the Kaufman (broad area), electron impact, duoplasmatron, surface ionisation and liquid metal ion sources. A useful review is given by R. Smith and J. M. Walls in *Methods of Surface Ananlysis,* J. M. Walls (ed), pp 20-30, Cambridge University Press, 1990.

ion beam sputter cleaning. A surface treatment whereby atoms from the surface of an object are removed by the sputtering action of an impinging high energy ion beam. The efficiency of this process depends on ion mass, ion energy, angle of incidence of the ions, physicochemical state of the surface, and temperature. Argon ions are usually preferred. Ion beam sputter cleaning is widely used as a means of preparing sample surfaces for interrogation by surface analytical methods, especially AES, XPS and SIMS.

ion beam sputter deposition. A sputter deposition process in which (usually) argon ions, generated in an ion beam gun, are used to remove atoms from a target surface by the mechanism of sputtering. The sputtered atoms recombine at a substrate surface and form a coating. This process falls within the plasma assisted PVD classification and permits the sputter deposition of magnetic and dielectric materials. It is usually performed at a lower pressure than magnetron sputtering.

ion bombardment heat treatment. See *plasma heat treatment.*

ion boriding. See *plasma boriding.*

ion carburising. Another term for *plasma carburising*, a term first used by Klockner Ionon for their plasma carbursing process.

ion cleaning. See *ion beam sputter cleaning*

ion current. The quantity of electrical current transported by ions in any given situation, e.g., in a glow discharge plasma.

ion implantation. A method of modifying surface microstructure and properties, whereby high energy ions (~10^{17}–10^{18} ions/cm^2) are implanted into the surface and sub-surface of a material, to a depth that increases in proportion with the incident ion energy. Nitrogen is the most popular ion, but a wide variety of metallic and non-metallic ions can also be implanted. The technique causes a marked hardening of metallic surfaces, and, depending upon the implanted element, some improvement in corrosive-wear resistance is observed, e.g., the Ti–6Al–4V shows improved corrosive-wear resistance in saline environments after implantation with 3.5×10^{17} ions/cm^2 of nitrogen (equating to a concentration of ~ 20 at.%). Chromium implantation has also been observed to improve the corrosion resistance of plain carbon steels.

The main limitation of the conventional ion implantation method is its highly directional nature and the shallowness of the treatment (up to 0.1 μm). Both these aspects are now being addressed by the newer approach of *plasma immersion ion implantation.* However, conventional ion implantation has its uses. It is widely applied for doping silicon wafer devices; it is also employed for hardening high precision tool steel moulds and dies, where the small dimensional distortion associated with other processes, like nitriding, cannot be tolerated. In the latter example wear rates must be completely curtailed by the implantation, to justify its use.

ion implanter. A device used to achieve ion implantation. The conventional design comprises a vacuum system, ion source, mass separator and linear accelerator. Objects are placed on a movable platform or multi-axis work station and are passed under the ion beam, which provides doses of $\sim 10^{17}$-10^{18} ions/cm^2.

ionised cluster beam PVD. An evaporative PVD process in which metallic vapour is passed through an orifice plate of such design that clusters of evaporant, rather than individual atoms, are created. These are then singly or multiply ionised prior to deposition.

Ionit®. The registered trade mark of Klockner Ionon (now Metaplas Ionon GmbH), for their plasma nitriding process.

ion nitriding. The original term for plasma nitriding first used by Klockner Ionon; the term has been superseded by *plasma nitriding,* but persists, especially in the United States.

ion plating. Also referred to as *evaporative source PVD*. One of the more important industrial PVD methods. It can be configured for the production of metallic (see *Ivadising*) and ceramic coatings (see *reactive ion plating)*. The process is conducted within a high vacuum vessel (initially evacuated to better than 2 x 10^{-6} torr). Typically a negative DC voltage ($\approx$400-1000volts) is applied to the substrates while the chamber pressure is back-filled with a partial pressure of $\approx 10^{-2}$ torr of argon causing a glow dicharge plasma to surround the substrates. For reactive deposition, argon is partially replaced by, for example nitrogen, to produce nitride coatings. The metallic vapour phase is supplied by one or more evaporation source(s); these include arc source, electron beam source or hollow cathode source evaporators. A useful review of ion plating methods is given in A. Matthews, *Surf. Eng.*, 1985, **1**, (2), pp 93-104. Also see *arc source PVD, electron beam evaporative PVD, hollow cathode evaporative source PVD* and *triode ion plating.*

ion siliconising. See *plasma siliconising.*

ion sputter cleaning. See *sputter cleaning.*

ion titanising. See *plasma titanising.*

ion vapour deposition. See *Ivadising.*

Ivadising. A proprietry name (circa 1974) used by the McDonnel Aircraft Company (USA) for their ion plating system used for coating aerospace components (principally steel fasteners) with aluminium for corrosion protection. The technique was devloped to replace cadmium plating which was banned because of its high toxicity. Three common coating thicknesses are used: 7.5 µm, 12.5 µm and 25 µm. The thicker coatings provide the greatest longevity of corrosion resistance. Thin coatings are used where compliance to small tolerances is required, e.g., on the threads of certain fasteners. Following deposition, it is usual practice to glass bead peen the coatings. This helps close micropores and provides a smoother surface finish. At this juncture parts with poorly adhered coatings can be identified and scrapped. Apart from steel, numerous titanium and aluminium alloy aircraft components are regularly coated. These include electrical connectors, fuel and pneumatic line fittings and forged engine mountings. Also see *ion plating.*

IVD. Ion vapour deposition; another name for Ivadising.

J

Jet-Kote. See *high velocity oxygen fuel (HVOF) spraying.*

jet vapour deposition (JVD). A relatively new method for making thin, high density coatings under reduced atmosphere, originally devised by Halpern and Schmidt in the 1980s. The process comprises one or more metallic vapour jet source(s), housed within the confines of a vacuum chamber. By maintaining a lower pressure within the jet source with respect to the chamber, metal vapour is ejected at high speed, through a narrow orifice, towards the substrate. The method is capable of producing metal or ceramic coatings, the latter being achieved via reactive deposition. For a recent paper see A. R. Srivatsa, D. T. McAvoy, D. L. Johnson, J. J. Schmidt and B. L. Halpern in 'Surface Modification Technologies VIII', 1995, London, The Institute of Materials.

K

keying-in. The aspect of achieving coating adherence through mechanical interlocking. A surface maybe deliberatley roughened, for example, via grit blasting, to achieve this effect.This effect can be seen at the coat-substrate interface in the micrograph shown in the ***colour section,*** p. C.

Knoop hardness. A method of microindentation hardness also termed toucan indentation hardness. The diamond indenter has a diamond shaped pyramid, with one apex (along the major axis) at 172.5°and the other (along the minor axis) at 130°. Knoop hardness is given by the equation:

$$H_K = 2P/L^2 \text{ (cot } 172.5° + \tan 130°) = \text{kg/mm}^2$$

$$\text{or } H_K = 14.23\ P/L^2 = \text{kg/mm}^2$$

where L is the long diagonal length and P is the load in kg.

L

lacquering. A procedure widely used in the electroplating sector for the tarnish protection of electroplated coatings (such as brass) and electro- or mechanically polished objects. Tinted lacquers are also used to impart a decorative finish. Numerous compositions exist but there are two broad catagories of lacquer: (i) air drying lacquers and; (ii) stoving lacquers. Air drying lacquers can be based on nitrocellulose or synthetic resins like vinyls and acrylics. Stoving lacquers are invariably based on synthetic resins and include urea-formaldehyde, melamine-formaldehyde, epoxies and acrylics. Three established methods of application are: brushing, dipping and spraying. A recent refinement has been the use of an electrostatic spray method. This has superior throwing power for external features. Unfortunately, internal features cannot be treated by the electrostatic action. Interior surfaces can be sprayed by the direct assistance of compressed air but no electrostatic action is involved.

LACVD. See *laser assisted CVD.*

laser (Light Amplification by Stimulated Emission of Radiation). A laser is a device which achieves a finely controlled emission of coherent, monochromatic high energy electromagnetic radiation ranging from the ultra-violet to the infra-red wavelength range. The most common medium in which the lasing action can be created are gases of various compositions; in surface engineering the CO_2 laser is the most popular. The gas used comprises a mixture of 10% CO_2, 30% N_2 and 60% He which is rapidly circulated at reduced pressure between two pairs of electrodes which results in the generation of an electrical discharge and excitation of CO_2 molecules; when the excited electrons return to their lower energy levels photons are emitted which stimulate further photon emissions (from adjacent molecules) which are perfectly in phase and of the same wavelength (10.6 μm for CO_2 molecules). The emergent laser light has high phase contrast, can be focused to very small diameters, approximating to the wavelength of laser light, which enables high power densities (~10^3 to 10^{10} W/cm^2) to be realised. When coupled with appropriate substrate and/or laser beam manipulation, a wide range of interaction time, from 1 to 10^{-8} seconds can be obtained. By varying interaction time and power density a broad variety of laser surface engineering techniques are feasible (diagram).

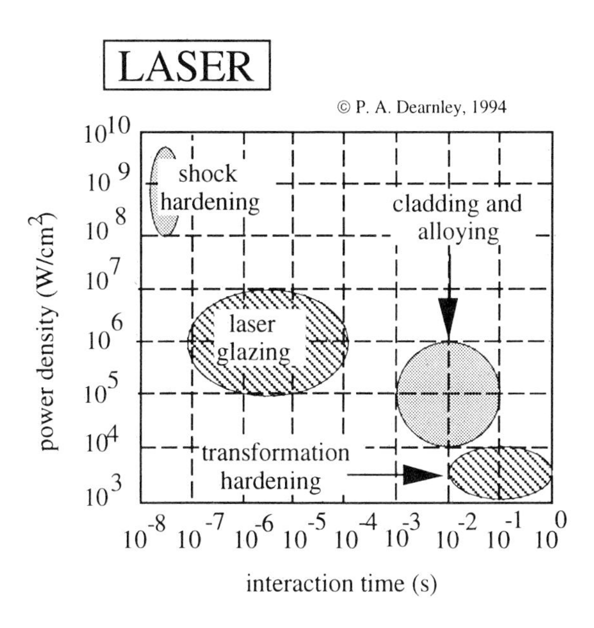

laser ablation. See *ablation.*

laser alloyed zone (LAZ). The surface zone whose composition has been modified by laser alloying. Typically between 20 μm and 1.00 mm deep.

laser alloying. See *power beam surface alloying.*

laser amorphisation. The same as *laser glazing.*

laser assisted CVD (LACVD). CVD in which appropriate reactive gases are passed over a laser heated surface. A plasma is formed above the sample surface, which, it is claimed, results in the favourable enhancement of CVD reaction kinetics.

laser beam deposition. See *laser evaporation.*

laser beam evaporation. See *laser evaporation.*

laser beam gas alloying. See *power beam surface alloying*

laser beam glazing. See *laser glazing.*

laser cladding. Laser treatment applied to a metallic surface which is momentarily surface melted to enable the encapsulation of preplaced or injected ceramic powder particles. The treatment is applicable to ferrous or non-ferrous metals and alloys. There has been a lot of interest in recent years in applying such treatments to aluminium and titanium alloys because both of these suffer from the disadvantage of responding poorly to thermochemical diffusion treatments, i.e., they develop very shallow treatments, if any. Laser cladding offers the possibility to develop deeply hardened surfaces, up to 1 mm. The process is analogous to *electron beam cladding.*

laser CVD. See *laser assisted CVD.*

laser evaporation. A vacuum evaporation (PVD) method in which a laser beam is scanned over the surface of a solid source or target, causing evaporation through laser ablation. The laser beam is passed through a glass port in the vacuum chamber. The process can be carried out under high vacuum conditions ($\sim 10^{-6}$ Torr) when the substrate remains at near ambient temperature, or the substrates may be negatively biased in an appropriate atmosphere (e.g., Ar) held at $\sim 10^{-2}$ torr which creates a glow discharge plasma, providing substrate heating (to temperatures ~400°C) and improving throwing power. One advantage of using a laser is that electrically insulating materials can be evaporated; this is less easy with the more usual electron beam methods. See P. A. Dearnley and K. Andersson, *J. Mater. Sci.,* 1987, **22**, 679-682.

laser gas alloying. Liquid phase alloying with an externally applied gas. See *power beam surface alloying.*

laser glazing. Also termed laser vitrification. Achieved by surface melting with a laser beam at power densities ~ 10^5 to 10^7 W/cm^2 for ~ 10^{-4} to 10^{-7} seconds; cooling rates exceeding 10^5 K/s can be realised which suppress the usual nucleation and crystallisation processes that accompany solidification. Instead, amorphous or vitrified (glass) surface layers are produced. The Nd–YAG or Excimer lasers are especially suited for laser glazing. The latter has recently been demonstrated to improve the statistics of fracture (i.e., the Weibull modulus) of Al_2O_3 ceramics by healing surface defects. This represents a significant breakthrough in materials processing. See E. Schubert and H. W. Bergmann, *Surf. Eng.,* 1993, **9**, (1), 77-81.

laser hardening

> 'Hardening using a high-energy laser as the heat source in a scanning or pulse mode' – IFHT DEFINITION.

Any laser treatment which results in hardening, in the absence of externally applied alloying elements or second phase particles. Hardening often results from the rapid solidification following laser melting, i.e., due to the creation of a very fine dendritic microstructure. In some cases crystallisation is suppressed and a hard glass is produced (*laser glazing*). Laser hardening can also be produced by manipulation of solid state transformations; see *laser transformation hardening.*

laser healing. A method of eliminating micropores in thermally sprayed coatings, through the action of laser melting. It has shown promise for healing pores in plasma sprayed MCrAlY coatings. For example see: R. Sivakumar and B. L. Mordike, *Surf. .Eng.*, 1987, **3**, 299-309.

laser heat treatment. Any surface modification resulting from the action of laser beam heating.

laser impulse. See *pulse mode laser.*

laser ionisation mass analysis. See *LIMA.*

laser irradiation. The interaction of a laser beam with any surface.

laser melting. Melting a surface through laser irradiation. The depth and shape of the melt zone is dependent upon the incident power density and laser interaction time.

laser nitriding. Liquid phase alloying with nitrogen gas. See *power beam surface alloying*

laser particle injection. See *power beam surface alloying.*

laser processing. Any laser assisted process. A very broad term encompassing laser heat treatment, laser welding and laser cutting.

laser pulse. See *pulse mode laser.*

laser substance. Any solid, liquid or gaseous substance which serves as the laser active medium. See *laser*

laser surface alloying. See *power beam surface alloying*

laser surface modification. Any surface modification resulting from the action of laser beam heating.

laser transformation hardening. A process applied to ferrous alloys with >0.35wt-%carbon. Laser power densities ($\sim 10^3$ to 10^4 W/cm^2) and interaction times ($\sim 10^{-2}$ to 1 s) are adjusted to cause rapid austenitisation of the surface; the heated zone subsequently self-quenches by conductive heat loss into the steel core beneath (water quenching is not required) which remains at ambient temperature. The resulting martensitic case can be furnace tempered, but this is not usually practised. The process produces less distortion than, for example, induction hardening, since only a very small volume of the object is heat treated.

laser treatment. Any laser assisted process. A very broad term encompassing laser heat treatment, laser welding and laser cutting.

laser vitrification. See *laser glazing.*

Laws of friction. There are three 'Laws of Sliding Friction'. Although the first two have been attributed to Amontons (1699), Leonardo da Vinci described them much earlier (circa 1500). The third law is attributed to Coloumb (1785). The laws are:

(i) the friction force is directly proportional to the normal force.

(ii) the friction force is independent of the apparent sliding contact area

(iii) the friction force is independent of sliding speed

It should be noted that polymers do not follow the first two laws because of their tendancy to flow plastically at their surfaces and to undergo localised asperity welding (seizure or adhesion) even under very low contact loads. For an excellent review consultant the work of Hutchings given in the Bibliography.

layered coating. Any multi-layered coating.

lead plating. Electroplated lead can be applied to steel surfaces with greater uniformity and density than is possible by hot-dipping; it is also more ductile. An acidic fluoroborate electrolyte is used in a rubber or plastic lined tank; anodes are 'chemically pure' lead. Deposition is carried out using a cathode current density of 1.5 to 2.0 A/dm^2 at 1.5 to 2.0 volts. The coating provides excellent protection from attack from mineral acids and is widely applied to the surfaces of bearing shells used in automotive engines, such as at the big-end and main-end. Also see *bearing shells* and *indium plating.*

lead-tin electroplating. See *tin-lead plating.*

leveller. (i) A substance added to an electroplating bath in order to produce a levelling or smoothing action on the electroplated deposit; (ii) an electrodeposit that is used to close surface defects prior to electrodeposition with the finish coating, e.g., electrodeposited copper serves in this manner for chromium-nickel electroplatings.

levelling action. The ability of a plating bath to produce a surface that is smoother than that of the substrate.

LIMA. Laser ionisation mass analysis. A laser (wavelength 0.26 μm) focused on a surface held under ultra high vacuum conditions, is used to vaporise a small volume of sample surface, previously imaged by light microscopy; the volatalised material is collected by a time-of-flight mass spectrometer and the mass to charge ratio of ions determined. Sampling is accomplished within a millisecond time interval. Spatial resolution is in the range of 1-5μm, while the analysed depth is ~0.5 μm; the latter can be reduced further (to ~30 nm) by defocusing the laser.

limiting current density. A current density which cannot be appreciably changed by causing a change in polarisation.

linear polarisation. The linear relationship between overpotential and current density which is regarded as prevailing at potentials very near to the corrosion potential.

liquid nitriding.

'Nitriding carried out in a liquid medium' – IFHT DEFINITION.

(i) A term for a specialist salt bath nitriding method. The initial bath composition is similar to that given under *salt bath nitriding/nitrocarburising*, but the ageing step is kept to a minimum thereby producing a bath of low NaCNO content and hence low nitrogen potential. It was mainly used for nitriding high speed steel taps and dies, but nowadays is infrequently used; (ii) a North American term for those processes described under *salt bath nitriding/nitrocarburising.*

liquid phase aluminising. See *salt bath aluminising.*

liquid phase boriding. See *salt bath boriding.*

liquid phase calorising. See *salt bath aluminising.*

liquid phase carburising. See *salt bath carburising.*

liquid phase chromising. See *salt bath chromising.*

liquid phase cyaniding. See *carbonitriding* and *salt bath carburising.*

liquid phase nitriding. See *liquid nitriding* and *salt bath nitriding*

liquid phase sherardising. See *hot dip glavanising.*

liquid phase siliconising. See *salt bath siliconising.*

localised carburising

'Carburising of only certain portions of the surface of an object'
– IFHT DEFINITION.

Carburising restricted to certain parts of a surface. Best achieved by *induction hardening* or *laser transformation hardening*. However, *paste carburising* can achieve similar results. Also see *stopping off.*

localised nitriding

'Nitriding of only certain portions of the surface of an object'
– IFHT DEFINITION.

Also see *stopping off.*

low-friction coating. See *antifriction coating.*

low-load hardness. See *nanoindentation hardness.*

low pressure carburising. See *vacuum carburising.*

low pressure CVD. Chemical vapour deposition performed below atmospheric pressure, typically ~10^{-2} to 100 torr. Also see *plasma assisted CVD.*

low pressure plasma spraying (LPPS). See *vacuum plasma spraying (VPS).*

LPCVD. See *low pressure CVD.*

LSA. Laser surface alloying; see *power beam surface alloying*

lubricant. Any fluid that serves to provide lubrication between solid sliding surfaces. Mineral oils of varying viscosity and additive content presently constitute the majority of lubricants used in, for example, automotive engines. The safe operating range of such oils is schematically presented in the diagram (adapted from R. G. Baker, 'Bearings in the automobile - a challenge for the materials engineer', *Metals and Materials,* **1**, (1), pp 45-52, 1985). Additives, like boundary lubricants and anti-oxidants, are used to broaden the operational range of such oils.

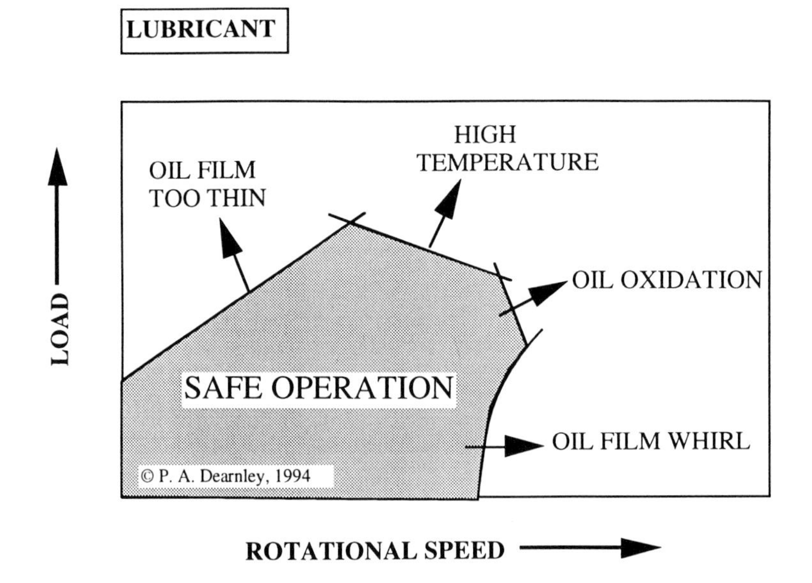

lubrication. A method of lowering the coefficient of friction between two sliding surfaces, achieved by separating them with: (i) a liquid lubricant (e.g., a mineral oil) and/or; (ii) a dry lubricous coating (e.g., molybdenum disulphide or lead). Also see *boundary lubrication, hydrodynamic lubrication* and *elastohydrodynamic lubrication.*

M

machining charts. A diagram depicting the operational cutting speed and feed rate ranges of a specific tool material used for turn-cutting a specific workpiece material. They also indicate the operational stability ranges for various types of chip morphology, e.g., built-up-edge and continuous chip forms together with the dominant tool wear mechanisms. Such diagrams were conceived by Trent (circa 1965) and were originally worked out for cemented carbide type cutting tool materials.They provide a practical guide-line for the optimal operating ranges of various tools. The diagram shown is for a steel cutting grade cemented carbide used on austenitic stainless steel. The optimal range resides between the shaded zones. At low speed where built-up-edge forms, carbide tools fail rapidly through a process of attrition. At higher speeds, failure is caused through plastic deformation of the cutting edge. Similar regimes exist for other workpiece materials, but the precise cutting speeds and feed rates differ significantly; this is due to widely differing variations in tool temperatures and stresses when cutting dissimilar materials.

machinability. The ease of machining any particular workpiece material with respect to the wear it induces on the recommended tool material(s). A workpiece of high machinability is one that causes minimal tool wear and consequently maximum tool life. The requirement by manufacturing industry for high machinability steels resulted in the design of 'free machining' steels that contain relatively large quantities of manganese sulphides and/or lead inclusions which lower the energy of chip formation and hence reduce cutting tool tempertures and wear. For further details consult 'Metal Cutting' by E. M. Trent, 3rd edn, Butterworths, London, 1991.

MACHINING CHART - austenitic stainless steel

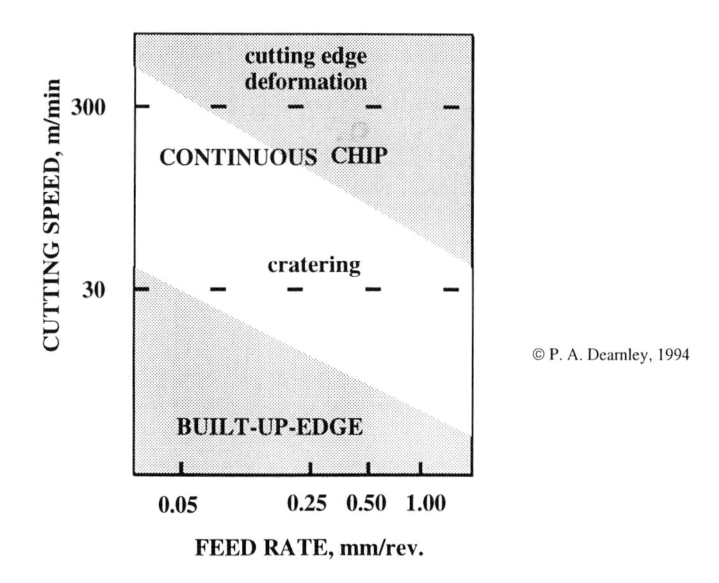

macroparticles. Particles deposited onto the surfaces of plasma assisted PVD coatings which eminate from an arc or electron beam evaporator. They can act as a source of failure in wear situations. See *filtered arc evaporator* and *shielded arc evaporator*

macrostress. *See residual stress.*

magnetron sputtering. A high rate sputter deposition (plasma assisted PVD) method. A tunnel shaped magnetic field is created in front of the target face; this serves to trap the available electrons in tightly defined orbits. The number of electron-atom and electron-molecule collisions in front of the target face is increased resulting in the creation of a high ion concentration. Hence, for a given pressure, more ions are available for sputtering through ionic bombardment than for a non-magnetron cathode; there is a marked increase in sputter yield and deposition rate. The following table shows that an order of magnitude increase in deposition rate is possible with magnetron enhancement.

The effect of magnetic confinement on sputter deposition rate

COATING	SPUTTER DEPOSITION RATE, μm/hr	
	BIAS SPUTTERING (no magnetic confinement)	**MAGNETRON SPUTTERING** (with magnetic confinement)
TiN	0.9	12.6
TiC	1.0	14.4
Ti(C,N)	0.9	14.4
Cr	0.5	28.8

manganising.
'Diffusion metallising with manganese' – IFHT DEFINITION.

Thermochemical treatment involving the diffusional enrichment of ferrous surfaces with manganese. The aim of the process is to increase the wear resistance of steel or cast iron objects according to the principle of high manganese (Hadfields) steel; a surface layer of retained austenite is developed, which on application of an external stress, may partly or completely transform to martensite. Manganising is usually carried out as a pack method at 1000°C for 6 hours. The pack comprises powdered manganese (source), Al_2O_3 (diluent) and NH_4Cl. (activator). The thickness of the layer produced does not exceed 25 μm. It is rarely practised outside Eastern Europe.

manual hardening. See *flame hardening*.

masking. See *stopping off*.

mass transfer media. The four states of matter used to transfer mass, i.e., solids, liquids, gases and plasmas. In surface engineering these usually refer to pack, salt-bath, gaseous and plasma assisted methods.

matt surface finish. A surface of poor specular reflectivity. Also see *reflectance.*

mechanical cleaning. Any process that renders surface cleaning by mechanical action.

mechanical coating. See mechanical plating.

mechanical galvanising. A mechanical plating process in which a zinc coating not greater than 25 μm is produced.

mechanical plating. A method of coating whereby a combination of mechanical and chemical action is used to produce a coating from powdered metal. The process is carried out in a ball mill. Also called peen plating.

mechanical polishing. Polishing performed by pressing a surface against a revolving soft wheel impreganted with fine (<10 μm) abrasive particles such as diamond or Al_2O_3.

mechanical surface engineering. A general term embracing the treatments aimed at modifying the properties of the surface of an object using continuous or dynamic mechanical action. The main types of mechanical surface engineering are: *shot peening, mechanical plating, roll hardening* and *friction surfacing.*

melt extrusion coating with plastics. A plastic coating process in which molten plastic is forced through a slotted die, with an adjustable opening, onto the object(s) being coated.

metallic lustre. The brilliant surface lustre of metallic surfaces, especially those free from surface oxide.

metallic painting. A paint containing metal powder. For example, aluminium powder is used to impart moderate corrosion protection for ferrous surfaces while paints containing copper or gold particles are used for decorative purposes.

metalliding. An electrolytic salt bath method devised at General Electric in the USA, circa 1962 (US Patent 3024176). One possibility is to use it for boriding, when the salts comprise a 9:1 mix of KF–LiF containing 10wt-% KBF_4 protected by an argon shroud. It has the virtue that lower operating temperatures (from 600 to 900°C) are possible, compared to those required for fused borax (see *salt bath boriding*). However, despite early enthusiasm, the process has not been adopted on a significant industrial scale.

metallising. See *diffusion metallising*. Also sometimes meaning those thermal spraying methods used for producing metallic coatings.

metal-organic CVD (MOCVD). Chemical vapour deposition from the reduction or pyrolysis of metal-organic compounds. Widely used in the electronics industry for producing GaAs films.

metal spraying. Thermal spray deposition of metallic coatings.

M_f temperature. The temperature at which austenite completely decomposes to martensite.

micro-cracked chromium plating. See *chromium plating*

microhardness. Hardness measured using diamond pyramide indenters (Vickers or Knoop method) at loads ranging from 2 g to 2 kg. Also see *hardness profile, Berkovich hardness, Vickers hardness* and *Knoop hardness.*

microporous chromium. See *chromium plating*

microstress. *See residual stress.*

microwave plasma CVD. Plasma assisted CVD in which the plasma is generated by microwaves, normally introduced into the reaction chamber from an external microwave generator. Typically, a frequency of more than 1000 MHz is used. This technique is presently popular for the synthesis of diamond films, using for example, 10%CO and 90%H_2, as the input gases; deposition temperatures of 750-885°C are typical.

minor thermochemical diffusion techniques. Over the course of this century it has been demonstrated that a very large number of elements, either individually, or in combination with one or two others (termed multicomponent systems), can be successfully introduced into the surfaces of (mainly) steels by thermochemical diffusion at elevated temperature. Accordingly, a diverse variety of compound formulations have been developed to make this possible by means of pack, paste, salt bath and gaseous methods. Those treatments of significant industrial importance have been given their own entry in this guide. The table is a list, although not exhaustive, of systems that have only received experimental investigation or are **not** practised on a significant industrial scale. Unless otherwise indicated, the treatments are applied to steels. Processing times listed in the table do not refer to the paste method; in such cases this ranges from a few minutes up to a maximum of 1 hour. With respect to salt bath vanadising the reader is also referred to the entry for *Toyota diffusion (TD) process*.

Summary of minor thermochemical diffusion techniques

NAME	DIFFUSING ELEMENTS	MASS TRANSFER MEDIA*	TEMP., °C	TIME, h	TREATMENT DEPTH, µm
alumino-siliconising	Al, Si	gas, pack, paste, salt bath (i)	900-1200	<10	<60
antimonising	Sb	pack, salt bath (i)	600-700	1	2-5
berylliumising#	Be	gas, pack, salt bath (e)	900-1200	3-6	<30
chrome-siliconising	Cr, Si	gas, pack, paste, salt bath (e, i)	980-1100	3-10	<100
chrome-alumino-siliconising†	Cr, Al, Si	paste	900-1400	1	50-120
copperising	Cu	gas, salt bath (e, i), pack	300-1025	3-5	<150
molybdenising#	Mo	paste, pack	900-1260	1-12	<150
niobiumising#	Nb	gas, salt bath (e)	1000-1300	1-6	<100
phosphorising	P	salt bath (e, i), pack	910-1150	6-8	<250
titanising	Ti	gas, pack, paste, salt bath (e)	800-1250	3-10	<70
tungstenising	W	gas, pack	900-1400	5-6	<100
vanadising	V	pack, salt bath (i)	900-1150	3-6	<250
zirco-aluminising	Zr, Al	pack	900-1100	2-8	<350

* (e) = electrolytic, (i) = immersion/electroless; # ferrous and non-ferrous alloys; † mainly applied to niobium

MO-CVD. See metal-organic CVD.

modulated coating. A thin coating (<20µm) comprising alternating layers of two or more distinctly different metals or ceramics, deposited by, for example, one of the plasma assisted PVD methods. The variation in composition is typically sinusoidal; hence, for a duplex variant, it is feasible to classify the coating in terms of a compositional wavelength, which can be as small as 8 nm. Also see *superlattice coating*.

modulated current plating. See *pulse plating*.

molybdenising. The aim of the process is to increase corrosion resistance of iron, steels and superalloys. See *minor thermochemical diffusion techniques.*

molybdenum disulphide. A popular coating used as a solid lubricant. Widely used in the internal mechanisms of satellites and space vehicles but also found in many less exotic applications.

***M_s* temperature.** The temperature at which austenite commences decomposition to martensite.

Multicomponent boriding systems with pack media

Multicomponent boriding technique	Pack composition, wt-%	Optimal process sequence*	Substrates	Treatment temperatures, °C
Boroaluminising	(i) 84% B_4C + 16% borax (ii) 97% ferroaluminium + 3% NH_4Cl	B->Al	plain carbon steels	1050
Borochromising	(i) 5% B_4C + 5% KBF_4 + 90% SiC (ii) 78% ferrochromium + 20%Al_2O_3 + 2% NH_4Cl	B->Cr	plain carbon steels	boride at 900 chromise at 1000
Borosiliconising	(i) 5% B_4C + 5% KBF_4 + 90% SiC (ii) 100% Si or 95% Si + 5% NH_4Cl	B->Si	0.4% C steel	900 to 1000
Borovanadising	(i) 5% B_4C + 5% KBF_4 + 90% SiC (ii) 60% ferrovanadium + 37% Al_2O_3 + 3% NH_4Cl	B->V	1.0%C steel	boride at 900 vanadise at 1000

*e.g., B->Al = boride then aluminise

multicomponent boriding. A general term whereby boron and one other element are diffused into the surface of a component. The elements maybe diffused consecutively or simultaneously. Such treatments are usually designated according to the elements involved, e.g., borochromising and borotitanising refer respectively to diffusion of B + Cr and B + Ti. Such terms do not give any indication of weather the elements are introduced consecutively or simultaneously. Much experimentation has revealed that, on the whole, treatments are most reproducible when boriding is carried out first, followed by diffusion with an appropriate metal (Cr, Ti, Al) or non-metal (usually Si). Such treatments have received greatest development in Eastern Europe where the prefered mass stransfer media is the pack method. Consecutive treatments involve carrying out two separate pack heat treatments, making process economics unattractive. However, the development of appropriate gaseous or plasma diffusion multicomponent boriding methods would be a more viable option. The table lists some common multicomponent boriding systems. Only pack compositions are given. It should be appreciated that these techniques can also be carried out using appropriate gases, salt baths, and, in the case of simultanous multi-component biriding, appropriate pastes. Other systems have also been investigated, e.g., B+P, B+Zr, B+Cu, B+W and B+Mo. Some of these treatments are detailed elsewhere in this guide; see *borocopperising, boromolybdenumising* and *borotungstentising*.

multicomponent thermochemical treatment. Thermochemical treatment involving the enrichment of the surface layer of an object simultaneously or consecutively with more than one element.

multi-layered coating. See *composite coating* and *modulated coating*.

multiple-plate system. A composite multi-layered coating produced by electroplating.

multi-stage nitriding. Another term for double-stage nitriding. See comments on Floe Process under *gaseous nitriding*.

N

nanoindentation hardness. A recent method of microhardness or ultramicrohardness using very small indentation loads ~10μN to 5mN in conjunction with a Berkovich (triangular base pyramid) diamond indenter. Machines record the penetration depth as a function of indenter displacement. The technique is able to determine the hardness of very thin coatings (<1μm thick); from the slope of the load-unload curves elastic moduli can also be derived. A disadvantage of the technique is its high sensitivity to sample surface roughness.

Nd-YAG laser. A solid-state laser in which an yttrium-aluminium-garnet (YAG) crystal, doped with neodymium (Nd) ions, constitutes the laser active medium. With this laser it is possible to achieve laser pulses of a very short duration (several ns) each being of extremely high power (up to 1 GW). Hence, this laser is ideally suited for *laser glazing*.

nickel electroplating. See *nickel plating*.

nickel-iron plating. A nickel electroplating process in which nickel is partly substituted by iron. The baths are less easy to control than those used for nickel plating. The coating primarily has a decorative function and is characterised by full brightness, high smoothness, excellent ductility and good receptivity for chromium. However, such coatings are rarely used.

nickel plating. An electroplating process in which nickel is deposited from a Watts, sulphamate or fluoborate bath. Nickel electroplating is applied for corrosion protection, primarily to steel but also brass and zinc. To a minor degree it is also used for decorative purposes. Also see *electroless nickel plating* and *Watts bath.*

niobiumising. The aim of this process is to increase the corrosion resistance of iron, steel, nickel and vanadium. See *minor thermochemical diffusion techniques.*

nitridability. The ability of a material, usually a steel, to increase its surface hardness by the solution of nitrogen. For steels, it is essential they contain elements that develop coherent nitride precipitates during nitriding. Typically, these elements are Cr, Mo, Ti, V and Al. Nitridability also refers to the rate of nitride case development. A steel of high nitridability is one which develops a nitrided case of significant thickness, at a given temperature, after a relatively short time.

nitrided case

> 'Surface layer of an object within which the nitrogen content has been increased by the nitriding process' – IFHT DEFINITION.

nitrided case depth

> 'Distance between the surface of a nitrided object and a limit characterising the thickness of the layer enriched in nitrogen. This limit should be specified' – IFHT DEFINITION.
>
> Also see *case depth.*

nitride layer. See *compound layer.*

nitrider. Jargon; usually referring to a gaseous or plasma nitriding vessel.

nitriding

> 'Thermochemical treatment involving the enrichment of the surface layer of an object with nitrogen' – IFHT DEFINITION.

Thermochemical diffusion treatment involving the enrichment of a metallic surface with nitrogen. Plasma, gaseous and salt bath media can be used. Due to its dependence on cyanide based salts, the latter is now less widely practised.

nitriding. *Metallographic section of a 722M24 low alloy steel, plasma niitrided in cracked* NH_3 *at 500° C for 10 h, showing: (i) the exterior* γ'*–*Fe_4N *compound layer; (ii) the adjacent diffusion zone. Light optical micrograph (etched in 2% nital).*

The process is most effective on prior hardened and tempered low alloy steels containing alloying additions of chromium, molybdenum and aluminium. Specific steels that can be nitrided are given in the table under *plasma nitriding* together with details of achievable case depths and surface hardness. For most steels, nitriding is conducted at temperatures between 400 and 590°C; hence, a ferritic treatment. Austenitic stainless steels can be readily plasma nitrided between 475 and 700°C, but gaseous nitriding requires a chemical pre-treatment to remove the passive layer. For low alloy steels (micrograph) strengthening is within the diffusion layer (constituting ≈ 99% of the case depth). This is achieved by the precipitation of coherent nitrides of chromium, molybdenum and sometimes aluminium. This places the case into a state of residual compressive stress that is very effective in opposing crack growth. Nitriding is therefore very effective in increasing the rolling contact fatigue endurance (and strength) of low alloy steel gears. In more recent years, nitriding has been extended to include titanium alloys. Here higher processing temperatures are needed (see *plasma nitriding*) and very little residual stress is developed. Case depths are shallower than for steels and wear protection is mainly provided by an external duplex layer of Ti_2N and TiN. The inner Ti_2N layer is approximately 5 times the thickness of the outer TiN and has a strong crystallographic orientation such that the {002} planes lie parallel to the surface. The outer TiN layer does not exhibit a strong crystallographic orientation. Underneath the duplex layer, the titanium alloy matrix is strengthened through solid solution hardening. This zone is relatively shallow, but does provide some support for the outer Ti_2N/TiN layer which is helpful in applications where Hertzian stresses develop. In this sense nitrided titanium alloys are superior to titanium alloys coated with TiN by conventional plasma assisted PVD processes. Certain cobalt alloys also lend themselves to nitriding, e.g., the stellite group.

nitriding response. See *nitridability.*

nitriding steels. Low alloy steels typically containing ~0.25 wt-%C and 3.0 wt-% Cr with minor additions of molybdenum (~0.50 wt%). Among the most popular types of nickel-chromium-molybdenum steels are BS970:722M24. Chromium can be reduced and replaced by aluminium. This serves to enable higher prior tempering tempertures, without impairing nitridability. (Note: chromium carbide precipitation is significant on tempering above 550°C; such chromium is not able to participate in subsequent nitriding reactions. Hence the virtue of aluminium bearing steels like BS970:905M31.)

nitrocarburised case

> 'Surface layer of a nitrocarburised object, consisting of a compound layer and a diffusion zone' – IFHT DEFINITION.

nitrocarburising

> 'Thermochemical treatment which is applied to a ferrous object in order to produce surface enrichment in nitrogen and carbon which form a compound layer. Beneath the compound layer there is a diffusion zone enriched in nitrogen' – IFHT DEFINITION.

A thermochemical diffusion treatment applied to plain carbon steels whilst in the ferritic state. Nitrogen (the major solute) and carbon (the minor solute) are simultaneously diffused into the steel surface at temperatures ≈550-590°C for times that are typically less than 3 hours. The treatment produces an external compound layer comprising mainly ε-$Fe_{2\text{-}3}N$ (~5-15µm thick) although some γ'-Fe_4N may also exist as a minor constituent. Small increases in fatigue strength and endurance can be achieved by developing compressive residual stresses in the diffusion zone beneath the compound layer. This is achieved by quenching the components into water or an oil/water emulsion after completion of the diffusion stage. Although first popularised as a cyanide based salt-bath technique, environmental legislation has lead to significant developments in gaseous (see *Nitrotec and Nitrotec S*) and plasma nitrocarburising methods. These are now replacing salt bath methods.

nitrogen austenite

> 'Solid solution of nitrogen in gamma iron' – IFHT DEFINITION.

nitrogen ferrite

'Solid solution of nitrogen in alpha iron' – IFHT DEFINITION.

nitrogen potential (*r*). In gas nitriding the nitrogen potential can be determined by the level of dissociation in the exhaust gas. Nitrogen is reduced from ammonia via the catalytic reaction:

$$NH_3 \longrightarrow 3/2H_2 + N_{(Fe)}$$

Hence, the amount of nascent nitrogen available for solution (the nitrogen potential) can be expressed by:-

$$r = pNH_3/[pH_2]^{3/2}$$

where pNH_3 and pH_2 are the partial pressures of ammonia and hydrogen respectively. It can be readily appreciated that the nitrogen potential, *r*, can be reduced by increasing the partial pressure of hydrogen. This effect is exploited to enable *bright nitriding*, or nitriding with minimal compound layer (γ'–Fe_4N and/or ε–$Fe_{2-3}N$) formation. Also see *gaseous nitriding*.

nitogen profile. Variation in nitrogen distribution as a function of depth below the surface.

Nitrotec. A proprietry nitrocarburising treatment for low carbon steels marketed by Lucas Nitrotec Services (UK). The process is a refinement of an earlier technique known as black nitrocarburising which was first developed for the hydraulics industry. Nitrotec comprises three main steps. First, components are nitrocarburised in an ammonia/endothermic gas mixture. Second, components are given a rapid or "flash" oxidation after which components maybe slowly cooled or if maximum strength is required, water quenched. Third, components are immersed in a hot organic wax mixed with corrosion inhibitors. The result is a carefully engineered surface. The external compound layer (mainly ε-$Fe_{2-3}N$) contains up to 10% iron oxide; of this the outermost 0.5 to 1.0μm is almost entirely Fe_3O_4 which gives the components an attractive matt black finish. Micropores within the compound layer serve to retain the organic wax. The combination of iron oxide and retained wax/ inhibitor serve to provide significant resistance to atmospheric and saline corrosion. The application of nitrotec to microalloyed steels enables the yield strength of thin sectioned (~0.5 mm) material to be increased from 150 to 750 MPa providing the steels are quenched from above 550°C.

Nitrotec C. A proprietry gaseous austenitic nitrocarburising treatment marketed by Lucas Nitrotec Services (UK). This differs from normal gaseous austenitic nitrocarburising in that the components are given a prior carburising treatment to raise the carbon content of the surface, thereby enabling a 'deep case' to be produced. This requires higher processing temperatures (~775°C) than conventional austenitic nitrocarburising, permitting hardening of the prior carburised zone on quenching. Despite the relatively high temperatures, a functional external compound layer of ε-$Fe_{2-3}N$ is produced.

Nitrotec S. A proprietry thermochemical diffusion treatment marketed by Lucas Nitrotec Services (UK) as an alternative to hard chromium plating of low carbon steels. A two stage heat treatment is used. The first step comprises a gaseous ferritic nitrocarburising followed by a gaseous oxidation. After slow cooling, components are mechanically polished to a mirror finish. The second heat treatment comprises a short duration gaseous oxidation. The finished components retain an attractive highly polished silver-black appearance.

non-crystalline structure. A structure lacking long range order; more usually termed glassy or amorphous.

non-electrolytic nickel plating. See *electroless nickel plating.*

non-stick coating. Any coating that inhibits sticking or fouling by solids or suspended solids. Apart from obvious domestic applications, such coatings are often applied to the internal surfaces of pipes used in the dairy and food processing industries for the conveyance of food products. Sometimes known as "release coatings". Also see *PTFE coating.*

normal pressure CVD. Chemical vapour deposition carried out at atmospheric pressure.

notch wear. A form of wear affecting metal cutting tools used in turning applications. Essentially it is an accelerated form of wear observed on the rake face, cutting edges and flank faces at the outer sides of the chip-tool contact. This wear has two definite mechanisms when cutting steels and cast irons: (i) oxidation and; (ii) microfracture. Uncoated cemented carbides and CVD coated carbides, with TiC as a major coating constituent, are very prone to the oxidative type of notch wear, especially when repeated cuts of the same depth-of-cut are being made. Al_2O_3 coatings appear to be most resistant, while TiN coatings are intermediate in their response. For further details see P. A. Dearnley and E. M. Trent, *Metals Technology*, 1982, **9**, 60-74.

NRA. Nuclear reaction analysis (or spectrometry). A very high energy (0.5-4.0 MeV) particle (H^+, D^+, $^3He^+$,$^4He^+$) beam is directed normal to the sample surface producing collisions with the lattice atoms of the sample. Some ions are back scattered and analysed (see *RBS*), while others cause the sample atoms to undergo nuclear reactions. Two types of nuclear reaction provide analytical data. The first is accompanied by an instantaneous emission of nuclear radiation; on analysis this yields both chemical and depth information. Hence, it can be used for non-destructive concentration-depth profiling. This method is sometimes termed "prompt radiation analysis". The maximum useful detection depth (below the surface) is ~5 μm. (Note: this is greater than for RBS.) The second type of analysis measures radio-activity emitted during the nuclear decay of isotopes created during initial particle irradiation. Measurements are carried out as quickly as possible after irradiation, since many reactions have half-lifes of only a few hours. This latter form of NRA, sometimes termed "activation analysis", is primarily used for the detection of impurities in bulk samples, i.e., where there are no surface concentration profiles. Both NRA methods are particularly suited to determining concentrations of low atomic number elements ($Z<14$) dispersed in high atomic number matrices, such as interstitial elements in metal matrices. Recently, prompt radiation analysis has proven useful for determining the depth-concentration profile of nitrogen in nitrided titanium. See: H. J. Brading, P. H. Morton, T. Bell and L. G. Earwaker, *Surf. Eng.*, 1992, **8**, 206-212.Also see *RBS* and *PIXE*.

nuclear reaction analysis. See *NRA*

nuclear reaction spectrometry. See *NRA*.

O

opacity. The property of some coatings not to transmit rays of a specified wavelength, e.g., the ability to absorb radio waves (radio-opaque).

optical coatings. Coatings developed for enhancing the optical properties of glass. Such properties include refractive index, infra-red absorbtion and reflectance. The most common coating materials are vanadium oxide, aluminium oxide, tantalum oxide, silicon dioxide and silicon oxynitride. These maybe applied singly or as multi-layered, multi-phased coatings, usually using RF plasma assisted PVD or CVD methods.

optical properties. Properties of a coating material which characterise its behaviour in relation to the electromagnetic radiation of the optical range (i.e. visible, infrared and ultraviolet) incident upon it (e.g. reflectance, transmittance, index of refraction).

optical reflectivity. See *reflectance*.

organic coating. General term for any form of polymeric (plastic) coating.

overcarburising

> 'Surface carbon content in excess of the desired carbon content, produced during carburising' – IFHT DEFINITION.

overlay coating. Any coating deposited without chemical reaction with the substrate.

overpotential. Also called overvoltage. The displacement of the equilibrium (steady state) electrode potential needed to cause an electrochemical reaction to proceed at a given rate. A detailed explanation of overpotential is beyond the scope of this work. It should be appreciated, however, that there are three broad catagories of overpotential: activation overpotential, concentration overpotential and resistance overpotential. More detailed explanations exist in numerous treatise on metallic corrosion. For example see L.L. Shreir's 'Corrosion', cited in the Bibliography.

overvoltage. See *overpotential.*

oxidation. Thermochemical diffusion treatment in which oxygen is deliberately introduced into the surface layer of an object. The process is carried out in a controlled atmosphere, or an air vented furnace. Some recent innovations in this area have been the thermal oxidation of zirconium which results in an external layer of zirconia which is highly wear resistant; further it is mechanically supported by a zone of oxygen solutionised zirconium beneath, making it tolerant to point contact loads. This material is presently being investigated for use in the bearing surfaces of orthopaedic implant materials, e.g., by Smith & Nephew Richards of Memphis, USA. Also see *steam treatment* and the *nitrotec* group of processes where selective oxidation plays an important role.

oxidational wear. A process observed in rolling or sliding contacts whereby oxygen ingresses the contacting surfaces leading to localised oxidation. The process is stimulated by the occurance of 'hot-spots'. The oxide grows to a critical thickness beyond which it is sheared off and removed. The kinetics of such dynamic oxidation are generally agreed to be more rapid than for static oxidation.The process can affect both coated (diagram) and uncoated surfaces. Many oxidational wear models exist, but the principal theory is attributed to Quinn. For example, see T. F. J. Quinn, *Wear,* 1992, **153**, 179-200. Also see ***colour section,*** p. B, which shows the oxidational wear of a TiN coated titanium alloy.

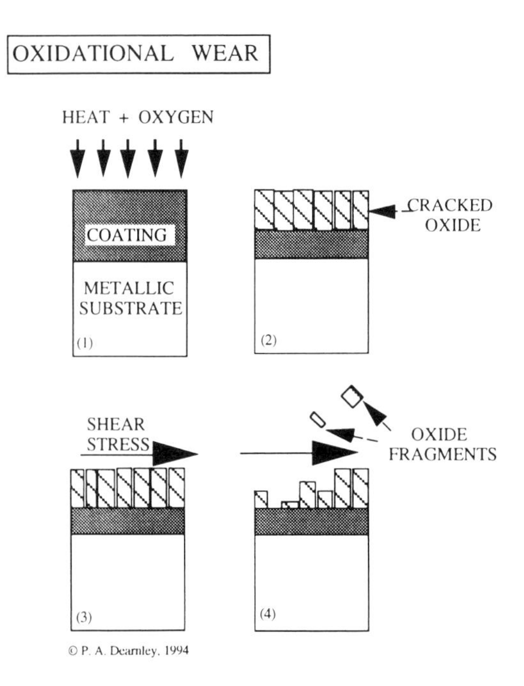

oxidation resistance. The ability of a material to withstand degradation by oxidising atmospheres, e.g., as found in gas turbine applications.

oxygen probe. See *zirconium oxygen sensor*

oxynitriding

'Nitriding involving the deliberate addition of oxygen'
– IFHT DEFINITION.

Thermochemical treatment involving the diffusional enrichment of a ferrous surface with nitrogen and oxygen. Oxynitriding has been applied to metal cutting tools made from high-speed steel. The process increases surface hardness; it is claimed that wear resistance is also improved. However, this is likely to be at the expense of toughness - a similar scenario takes place after nitriding high speed steel. Oxynitriding can be conducted at temperatures ≈520–560°C for 0.5–2 hours in a gaseous medium of ammonia and water vapour. The depth of the treated zone is in the range of 10 to 60 μm.

oxynitrocarburising. Nitrocarburising involving the deliberate addition of oxygen to the compound layer. Also see *Nitrotec* and *Nitrotec S*

P

pack aluminising. See *aluminising.*

pack boriding

'Boriding carried out in a powdered medium'
– IFHT DEFINITION.
See *boriding* and *pack cementation.*

pack boroaluminising. See *multicomponent boriding.*

pack borochromising. See *multicomponent boriding.*

pack borocopperising. See *borocopperising*

pack borosiliconising. See *multicomponent boriding.*

pack borotitanising. See *multicomponent boriding.*

pack calorising. See *aluminising.*

pack carburising

'Carburising carried out in a powdered medium'
– IFHT DEFINITION.

Maybe carried out using 100% hard wood charcoal, but more often, commercal carburising pack compositions comprise 6 to 20wt-% activating compounds bound to hard wood charcoal and/or coke by molasses, tar or oil. Barium carbonate has been a principal activator; due to its toxicity this is being discontinued in the United States because of industrial control legislation. Calcium, sodium and potassium carbonates are suitable alternatives. The activator(s) serve to create carbon dioxide via the reaction:

$$CaCO_3 \longrightarrow CaO + CO_2$$

which then reacts with the charcoal/coke:

$$CO_2 + C \Longleftrightarrow 2\,CO$$

The CO is then reduced to carbon at the steel surface where it is dissolved in the austenitic phase:-

$$2\,CO \longrightarrow C_{(Fe)} + CO_2$$

Treatment temperatures normally reside within the range of 815 to 1050°C, but there has been a recent trend to increase the upper limit to 1095°C. Like many other pack methods, the container holding the pack medium must be conditioned by a dummy run, such that its capacity to absorb carbon, at the expense of the components (in future cycles) is limited. Pack carburising is labour intensive and has the major disadvantage of requiring a second austenitising treatment to enable the mandatory quenching operation needed to develop the martensitic case.

pack cementation. (i) an obsolete term for pack carburising; (ii) Any pack method that results in the diffusional formation of a surface layer (or layers) enriched in interstitial or intermetallic compounds. Such processes include, pack boriding, pack aluminising and pack chromising. All these processes involve embedding components within a powdered pack contained inside a vessel, which is then placed inside a furnace. Sometimes inert or reducing gases are passed through the pack (e.g., pack aluminising and pack chromising) in other cases no external gases are required (e.g., pack boriding) although they still serve to prevent 'caking' of the pack and are recommended. Process control is difficult with such techniques, especially with regard to controlling the activity or potential of the diffusing specie(s). Pack constituents may be costly and are often used several times. After each diffusion cycle the powders become depleted in constituents and their activity or potential reduced. Industrial users therefore implement methods of "topping-up" powders with new constituents. This calls for stringent management of the powders. A detailed knowledge of the effect of various substrate compositions on pack life must also be accumulated. Also see *minor thermochemical diffusion techniques.*

pack chromaluminising. Carried out by embedding components in a mixture of powders comprising, Al, Cr, Cr_2O_3 (the sources of Al and Cr) and NH_4Cl. (the activator). The process is conducted at temperatures ~950–1100°C for 2–4 hours. The thickness of the layer produced does not exceed 100 µm.

pack chromising

'Chromising carried out in a powdered medium' – IFHT DEFINITION.

Conventional packs comprise a mixture of chromium metal or ferrochromium powder and an an activator such as NH_4Cl, NH_4Br or NH_4I. A gas tight container is required and hydrogen is passed through the pack which is heated between 850 and 1050°C for up to 12 hours. In a more recent development, components are packed in a mixture of $FeCl_3$ and metallic chromium powders, which on heating react to produce (probably) $CrCl_2$; argon or nitrogen is passed through the pack to prevent ingress of oxygen. $CrCl_2$ is reduced to Cr at the surface of the steel components. In both cases chromium subsequently diffuses into the steel and produces a surface layer whose constitution depends upon the steel substrate composition (see *Chromising*).

pack diffusion metallising

'Diffusion metallising carried out in a powdered medium' – IFHT DEFINITION.

Also see *minor thermochemical diffusion techniques, pack cementation, pack chromising, pack sherardising* and *pack siliconising.*

pack sherardising. Sherardising is carried out in a powdered medium consisting, as a rule, of zinc powder or dust with additions of zinc chloride or ammonium chloride to provide activation. Sometimes an inert substance (e.g. aluminium oxide, fireclay or high-silica sand) is added to act as a diluent and prevent sintering togther of the zinc particles. Temperatures ~400–800°C for 2–4 hours are typical. The process can be further assisted by rotation of the charge (tumbling). The thickness of the diffusion layer is ~ 30 to 200 μm. Also see *sherardising.*

pack siliconising.

'Siliconising carried out in a powdered medium' – IFHT DEFINITION.

Siliconising carried out by embedding components in a powdered mixture comprising silicon mixed with silicon carbide, silico-calcium or ferrosilicon. The diluent is usually alumina or magnesia, while NH_4Cl, NH_4I, NH_4F, KF or NaF act as the activator phase. The process requires ~4–6 hours at 950 to 1200°C.

PACVD. See *plasma assisted CVD (chemical vapour deposition).*

palladium plating. An electroplating process in which palladium is deposited from tetra-amino palladium nitrate solutions, which contain ~10-15g/l of palladium. The process uses platinum or platinised titanium anodes. Typical coating thicknesses are in the range of 2 to 5 μm. Palladium plating is used as a substitute for gold or rhodium plating, especially in the finishing of copper electrical contacts on circuit boards. It has a Vickers hardness of 200-300 kg/mm^2 which is higher than that of electroplated gold; hence palladium plating imparts reasonable wear resistance. It also finds use for the high temperature oxidation protection of refractory metals, such as tantalum and molybdenum.

PAPVD. See *plasma assisted PVD (physical vapour deposition.)*

partial pressure carburising. See *vacuum carburising.*

partial pressure control. See *closed loop partial-pressure control.*

paste boriding. Only practised when pack, salt bath, gaseous or plasma boriding is unfeasible or if only selected areas of a ferrous component needs to be borided. Pastes usually comprise a suspension of B_4C and cryolite (Na_3AlF_6) or B_4C-KBF_4-SiC in an organic binder such as nitrocellulose dissolved in butyl acetate or an aqueous solution of methyl cellulose. Commercial pastes are supplied in a viscous state and are water soluble, allowing a range of consistencies to be made, so that the paste may be brushed or sprayed on. After drying, another layer of paste is applied and the procedure repeated until a layer of dried paste some 1 to 2 mm thick is obtained. Components are then usually inductively heated, in a protective shroud of argon or nitrogen, to temperatures ~1000 to 1100°C for 10 to 20 minutes, enabling boride layers up to 50μm thick to be produced. A disadavantage of this method is that the resulting boride layers can be porous since the boron potential is higher than for other techniques. Paste boriding is principally confined to Eastern Europe and the states formerly constituting the USSR; it is rarely practised in the West. A number of multicomponent paste boriding methods exist. These enable the *simultaneous* diffusion of boron plus one or two other elements, e.g., paste boroaluminising, paste borochromising, paste borophosphorising, paste borotitanising and paste borozirconising.

paste carburising

'Carburising carried out with a surface coating in the form of a paste' – IFHT DEFINITION.

A rarely used thermochemical diffusion treatment. A steel surface can be selectively hardened by brushing with a paste comprising an organic binder, an activating compound and a carbon source. The latter two components can be potassium or barium carbonate and carbon lampblack respectievley. Diffusion of carbon is carried out at temperatures ≈ 910-1050°C for up to 4 hours, which is followed by water or oil quenching. The hardened surface layer is usually ~0.8-1.3mm deep.

PECVD. Plasma enhanced CVD. See *plasma assisted CVD.*

peening intensity. A characteristic of the shot peening process. See *Almen test.*

peen plating. See *mechanical plating.*

periodic reverse current (p.r.c). A form of electrical current used for electroforming or for obtaining thick electroplated deposits. An alternating current with an asymmetric wave form is used, where each cycle lasts several seconds. It is claimed that smoother electrodeposits are produced resulting from the selective dissolution of peaks during the reverse part of the cycle.

periodic reverse electrocleaning. Electrochemical cleaning in which the polarity of the object to be cleaned is deliberately alternated (every few seconds) between cathodic and anodic potentials.

phosphating. Iron and steel, zinc, cadmium and aluminium are the main metals that can be effectively phosphated. Zinc phosphate coatings are the most widely applied but others like iron phosphate and manganese phosphate are also in use. Phosphate coatings are generally classified as lightweight (1.5 to $4.5 g/m^2$), medium weight (4.5 to $7.5 g/m^2$) and heavyweight ($7.5 g/m^2$ minimum). The latter have superior load bearing capability and are frequently used as a solid lubricant on steel stock that is to be subsequently cold worked, e.g., by cold drawing, or cold forging. Medium weight coatings are used as lubricants for less severe cold working applications, while lightweight coatings are used as a corrosion protection pretreatment for subsequent coverage by organic paints. It should be noted that carburised, carbonitrided or nitrided steel surfaces cannot be satisfactorily phosphated. Phosphating treatments are commonly applied by immersion or spray techniques.

phosphorising. Thermochemical diffusion treatment involving the enrichment of steel surfaces with phosphorus. The aim of the process is to increase wear resistance and running-in ability of engaging steel components. Phosphorising can be applied by gaseous, pack and salt bath methods but is rarely practised outside Eastern Europe. The hardened layer is duplex comprising Fe_2P and Fe_3P with a Vickers hardness of 1100 and 1000 kg/mm^2 respectively. For a recent paper see J. Nowacki, *Wear,* 1994, **173**, 51-57.

physical adsorption. Surface adsorption in which the molecules of adsorbate and absorbent are held together by van der Waals bonds. Also see *chemisorption*.

physical vapour deposition (PVD). Any vacuum deposition process whereby one of the constituents of the final coating is vaporised or atomised from the solid state within the vacuum chamber, prior to deposition. The two methods of vapourisation/atomisation are evaporation and sputtering. In general plasma assisted processes like *ion plating* can be regarded as PVD, but for these cases a more exact definition is *plasma assisted PVD*.

PI^3. See *plasma immersion ion implantation*

pickling. See *acid descaling*

pin-on-disc test. A simple method of wear testing whereby a loaded pin slides against the face of a rotating disc under dry or lubricated conditions. Pin and/or disc wear maybe monitored. The US standard is ASTM G99.

pitting. Localised, but deeply penetrating, corrosion of a surface.

pitting index. A quantified measure of corrosion pitting resistance which enables direct comparisons between materials of differing composition or surface treatment.

PIXE. Particle induced X-ray emission; also sometimes called proton (H^+) induced X-ray emission. A very high energy (0.5-4.0 MeV) particle (usually H^+ or $^4He^+$) beam is directed normal to the sample surface. The ions collide with the lattice atoms of the sample. Some are back-scattered (see *RBS*) or trigger nuclear reactions (see *NRA*) while others cause K-shell ionisation, excitation and X-ray emission. The X-ray production cross section ($\sigma_{k\alpha}$) is related to the ionisation cross section (σ) through the fluorescence yield (ω_x) according to the equation:

$$\sigma_{k\alpha} = \sigma . \omega_x$$

The fluorescence yield is the probability of a radiative electron transition relative to non-radiative transitions. The magnitude of the X-ray cross section is inversely proportional to atomic number; the position of maximum X-ray cross section increases with increasing incident particle energy. The X-rays can be analysed by wavelength or energy dispersive methods. The advantage of using a high energy particle beam, rather than an electron beam (as in EDS or WDS) is that the signal-to-noise ratio is much more favourable, enabling better trace element sensitivity. The background 'noise' in WDS and EDS is due to the bremsstrahlung process; this is a continuous X-ray distribution associated with deceleration of the incident electron. From quantum analysis it can be demonstrated that the bremsstrahlung process decreases with increasing projectile mass; the effect is much less pronounced for H^+ or $^4He^+$ beam interactions, hence the more favourable signal-to-noise ratio is realised. PIXE has been carried out with micro-beams (~10 μm diameter) and satisfactory results can be obtained even in the absence of vacuum. The technique is still being developed for materials science applications.

plasma. See *glow discharge plasma* and *radio frequency glow discharge.*

plasma assisted CVD (chemical vapour deposition). CVD carried out at pressures ~1 to 10 torr. The kinetics of CVD are enhanced by the creation of a glow discharge plasma. The temperatures required to form carbides or nitrides by the reduction of metal halides (MCl_x) via the reactions:

$$MCl_x + H_2 + N_2 \rightarrow MN + xHCl$$

$$MCl_x + CH_4 \rightarrow MC + xHCl$$

are reduced from ~1000°C to ~450–550°C. When used to produce TiC coatings the technique is sometimes called ion or plasma titanising. Although a direct current (DC) glow discharge plasma maybe used, it is preferable to use a radio frequency (RF), microwave or pulsed plasma. Metal hydrides may also be used as the source of metal; see *pyrolytic CVD*. Plasma activation lowers the temperature required for pyrolytic CVD. Also see *radio frequency (RF) plasma CVD* and *microwave plasma CVD.*

plasma assisted PVD (physical vapour deposition). Any physical vapour deposition (PVD) process where a glow discharge plasma is used to assist with coating adherence or synthesis. For example see *magnetron sputter deposition* and *ion plating*.

plasma boriding. Also termed ion boriding or glow-discharge boriding. Like other boriding methods temperatures ~750 to 1000°C are used for the boriding of steels (hence, an austenitic treatment). In principle, the technique could be used for boriding non-ferrous metals like titanium, although this has not been explored. Hot or cold wall vessels can be used. The latter have the disadvantage of requiring a high power density DC glow discharge plasma leading to a very high boron potential. Hence, porous boride layers can be produced (See micrographs overleaf). The usual feed gas is a mixture of BCl_3 and hydrogen; the BCl_3 content should be less than 1% by volume. Pressures are ~5 torr. When using high power densities (required to achieve temperatures ~800°C with cold wall vessels) there is a high incidence of arc formation; this can be obviated by admitting up to ~50% Ar. The kinetics of plasma boriding are very similar to pack boriding. The advantage is that boron potential can be controlled in-situ and ought to assure the production of strain free mono-phased Fe_2B layers (see comments on residual stress covered under *boriding).*

plasma carbonitriding. Plasma carbonitriding is not presently industrially practised. In principle, this should be possible by developing the existing industrial plasma carburising capablity.

plasma carburising. Also termed ion carburising or glow-discharge carburising. It is carried out in the pressure range of 1 to 20 torr at temperatures ranging from 850 to 1050°C in a DC glow discharge produced in a mix of H_2 and CH_4 or just 100%CH_4. The load is cathodically charged and the plasma power density is kept below 5 W/cm^2 with the help of auxiliary radiant heaters. This assures uniformity of temperature while maintaining the carbon potential below that where soot formation might arise. It is usual practice to carry out a boost-diffuse cycle prior to quenching; such cycles represent a significant deviation from equilibrium conditions. For example, AISI 8620 steel maybe plasma carburised to produce a surface carbon of 1.1 wt-% after 1 hour at ~1000°C (the boost cycle). The methane supply is then turned off and the charge is kept at temperature for a further 1.25 hours in a neutral H_2 plasma. During this period (the diffuse cycle) carbon diffuses further into the component and the surface carbon content is allowed to fall to an optimal value of around 0.8 wt-%. The charge is then passed into an adjacent chamber and quenched in oil to develop the required martensitic case. Process economics dictate that conventional furnacing be used for the subsequent low temperature (~150-200°C) temper. There is some controversey regarding the economics of plasma carburising in comparison to gaseous carburising.

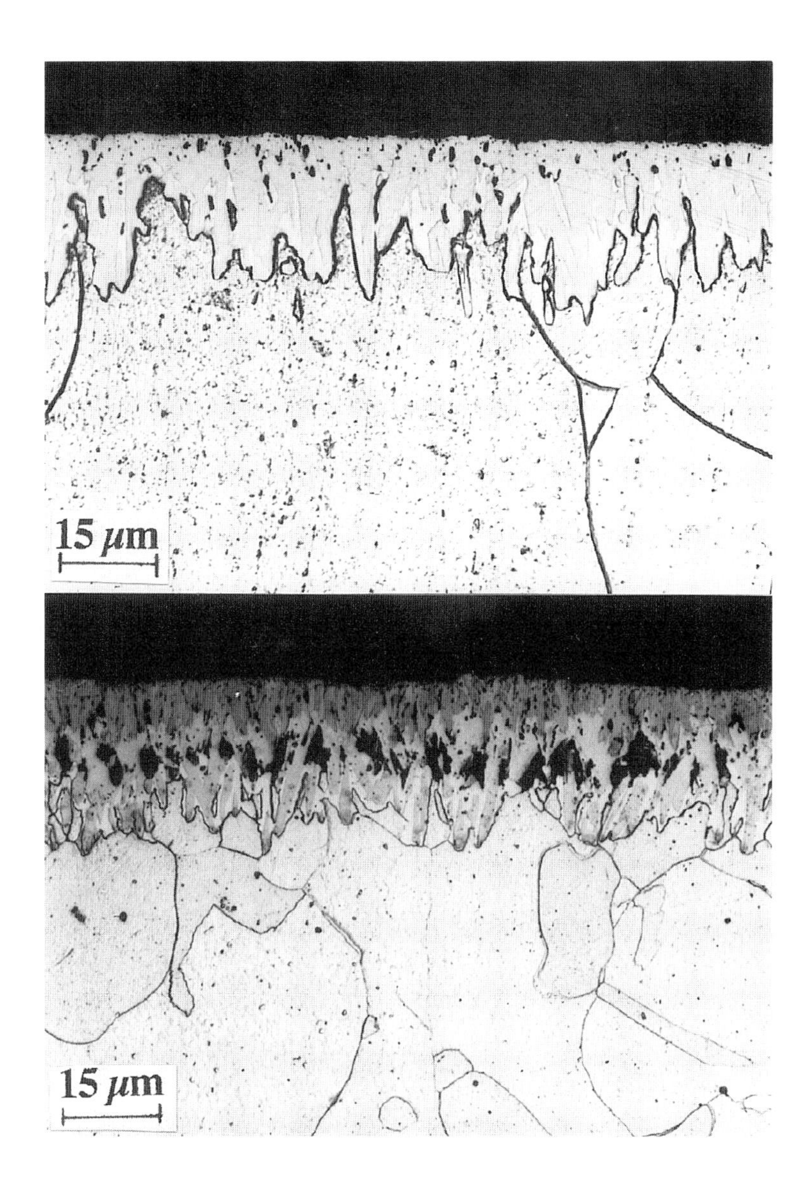

plasma boriding. *Metallographic sections prepared normal to the surface of a comercial purity iron following: (a) pack boriding; (b) plasma boriding. Although both samples were treated at 800°C for 2.5 h, the boron potential during the plasma treatment was much higher. This resulted from the high plasma power density required to heat the sample to 800°C; a cold wall vessell was used. The boride layer is not only duplex (FeB–Fe_2B) but contains Kirkendall porosity. This effect could be obviated by the use of radiant heaters; it would permit a lower plasma power density and hence a lower boron potential. This principle has been demonstrated in the related area of plasma carburising. Light optical micrographs (etched in 2% nital).*

plasma diffusion methods. Any process that utilises a glow discharge for the mass transfer of elements to a component surface. Such processes take several hours to complete to enable diffusion of the transported elements into the substrate.

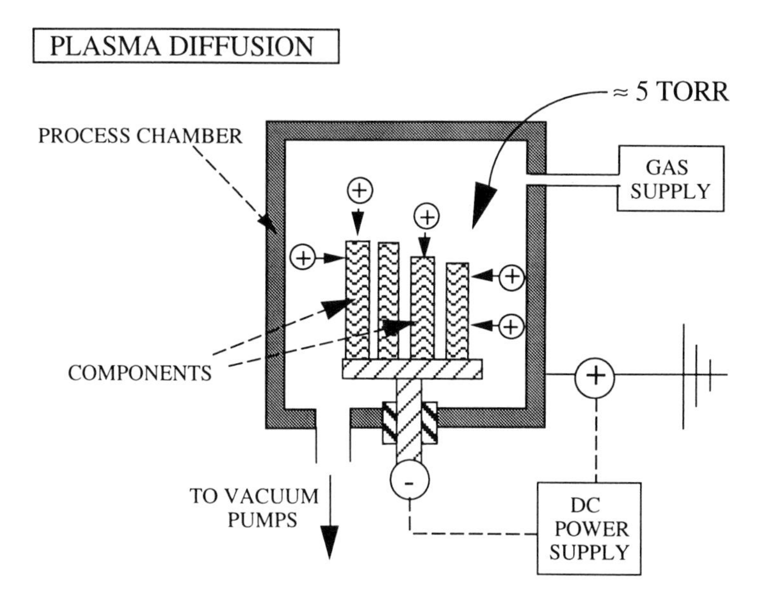

The components are heated by ionic bombardment; for processes requiring temperatures above 750°C, it is recommended that additional heat be supplied by radiant heaters. The latter procedure obviates the requirement of high power density plasmas (~5 to 10 W/cm^2) which are always in danger of collapsing into an arc discharge. Plasma diffusion methods include plasma nitriding, plasma nitrocarburising, plasma carburising and plasma boriding. Processing pressures are typically in the range of 1 to 15 torr.

plasma enhanced CVD. See *plasma assisted CVD (chemical vapour deposition).*

plasma etching. See *etching.*

plasma heat treatment

> 'Heat treatment utilising a current intensive glow discharge produced between the object to be treated (which acts as a cathode) and an anode in an appropriate atmosphere under reduced pressure'
> – IFHT DEFINITION.

A term encompassing all plasma diffusion methods and any annealing or hardening treatments where heat is provided (at least in part) by a glow discharge plasma.

plasma immersion ion implantation (PI^3). Also sometimes called plasma source ion implantation (PSII). A near non-directional method of ion implantation. A low power (~1 kW) radio frequency (RF) plasma (e.g., 13.56 M.Hz) is used to provide nitrogen ions which are accelerated towards test-pieces or components held at very high negative potential. The ions impact the surface with energies ranging from 20 to 200 keV. Since there is presently an interest in applying this treatment for improved wear resistance of steels, greater treatment depths are being produced by using higher treatment temperatures (~300-400°C) than used for conventional ion implantation. This necessitates the use of radiant heaters. Treatment depths up to 100μm can be achieved in this way. The process can be regarded as a form of nitriding, although quantification of the nitrogen potential remains a challenge. PI^3 is being developed at several research institutes throughout the world, however, its industrial viability has yet to be demonstrated. Its process economics require close scrutiny in comparison with plasma nitriding.

plasma metallising. Plasma spraying in which a metallic coating material is deposited.

plasma nitriding. Also termed ion nitriding or glow-discharge nitriding. First investigated and patented by Bernhard Berghaus between 1935 and 1945. Plasma nitriding is typically carried out at temperataures ~450 to 590°C for 10 to 64 hours, at pressures ranging from 1 to 10 torr in a direct current (DC) glow discharge plasma created in a $H_2 + N_2$ gas mixture. No auxiliary heating is required. The process requires a power density ~ $0.5 W/cm^2$ for cold wall vessels. There is some evidence that low power densities (below ~ $0.3 W/cm^2$) provide insufficient nitrogen potential to enable satisfactory nitride hardening. Hence, it may be advantageous to use cold wall chambers; plasma nitriding units utilising radiant heaters will tend to produce a lower operational plasma power density for a given processing temperature and load size. Nowadays very large production vessels are in regular use (see ***colour section*** pp. D and E), enabling charges ~1000 kg to be routinely processed.

As with other nitriding processes the highest surface hardness values are obtained with alloy steels containing nitride forming elements (especially chromium, aluminium and molybdenum). As the alloy content of a steel increases, nitrided case depth decreases for a given nitriding temperature and time (e.g., see *hardness profile*). Austenitic stainless steels can be nitrided without the need for chemical removal of the passive layer (this is removed by sputtering during the treatment cycle).

The following table provides an overview of the ranges of attainable surface hardness and case depths for a variety of DIN standard alloy steels. Although this data was obtained for plasma nitrided steels, it is also applicable for gaseous nitrided steels when utilising the appropriate nitrogen potential (stainless steels, however, require a special pre-treatment before gaseous nitriding). Typical hardness profiles are given under the citation for *hardness profile*.

Data for various plasma nitrided steels

DIN standard and steel type	AISI/SAE equivalent	Nitriding temperature (°C)	Surface hardness (HV/1kg)	Nitrided case depth (mm)
Structural				
St 37	1020	550-580	200-350	0.3-0.8
Nitriding				
31CrMo12		490-540	750-900	0.2-0.5
34CrAlNi7	A355	520-550	900-1100	0.2-0.5
High tensile				
42CrMo4	4140	500-550	550-650	0.3-0.5
30CrNiMo8	4340	490-540	600-700	0.3-0.5
30CrMoV9	::::	490-540	750-850	0.2-0.5
14CrMoV69	514	490-540	750-900	0.4-0.8
Hot work				
X32CrMoV33	H10	480-530	900-1100	0.1-0.3
X40CrMoV51	H13	480-530	900-1100	0.1-0.3
Stainless				
X5CrNi189	304	550-580	900-1200	0.05-0.10
X20Cr13	420	540-570	850-1050	0.1-0.2

It should also be appreciated that plasma nitriding has recently shown considerable potential for titanium alloys used in tribological applications, e.g., in the specialist custom car and motor racing sectors as well as specialised aerospace components. Here, because of a lower nitrogen diffusivity, higher nitriding temperatures (~800°C) are required than for steels. The treatment produces a duplex surface layer of TiN + Ti_2N supported by a nitrogen enriched α-Ti 'case'.

plasma siliconising. Also called ion siliconising or glow-discharge siliconising. A rarely practised industrial process. The components or test-pieces are heated between 900 and 1100°C for 1-2 hours in a direct-current glow discharge plasma containing silicon tetrachloride ($SiCl_4$) and hydrogen at 1-10 torr. The components are charged cathodically and are bombarded by silicon ions and partly dissociated $SiCl_4$ molecules. It has been claimed that plasma siliconising is kinetically more rapid than gaseous siliconising. Treatments up to 0.3 mm in depth have been produced on steels.

plasma source ion implantation (PSII). *See plasma immersion ion implantation (PI ^{3})*

plasma spraying. A thermal spraying process in which the coating material, in the form of powder, is introduced into a high enthalpy plasma torch, melted and propelled onto the surface of a component (particle velocities ~ 150 m/s), which maybe typically ~200 to 500 mm away from the torch exit. Both DC and RF plasma torches can be used; DC torches are more established. The process can be carried out in air at atmospheric pressure (here the technique is termed air plasma spraying (APS)) or at sub-atmospheric pressure in the presence of a protective atmosphere (see *vacuum plasma spraying*). It is a "line-of sight" process and works best on external surfaces. Special designs of plasma torch are available which enable spray coating of internal surfaces, but these are generally confined to relatively large ≥50 mm diameter bores.

Enthalpy of the plasma torch is controlled by: (i) changing the inter-electrode current; (ii) selection of the plasma supply gases. Diatomic gases, especially hydrogen and nitrogen, produce a higher enthalpy plasma torch than monotomic gases like argon and helium. Accordingly, plasma torch temperatures vary considerably; temperatures ~12000 K are easily attained. The creation of sound coatings, containing minimal porosity, is dependent upon achieving a high efficiency of powder melting. Hence, for high melting point ceramic powders, like alumina (Al_2O_3) and zirconia (ZrO_2), a high enthalpy hydrogen-argon gas mixture is required. The selection of a very high enthalpy plasma has to be balanced by the need to obtain satisfactory electrode life; this diminishes as the enthalpy of the plasma increases. Electrodes are usually made from thoria dispersed tungsten and are a significant cost factor. See ***colour section***, pp. C, F, G and H.

plasma titanising. See *plasma assisted CVD.*

plastic dip coating. A plastic coating process in which the objects to be coated are immersed in a well agitated polymer based liquid and subsequently withdrawn and drained. Solvent evaporation and curing finish the process. Plastic dip coating is useful for objects with complex shapes and inaccessible recesses.

plastic flow coating. A method of applying a plastic coating in which liquid resin is passed through nozzles of such design that all external component surfaces become covered; excess resin is drained, collected and recycled. Solvent evaporation and curing finish the process. Also see *curtain coating.*

plastic powder coating. See *electrostatic fluidised-bed coating* and *fluidised-bed coating with plastics.*

plastic roller coating. A process in which plastic is applied to a material surface, in sheet form, by rotating rolls. The polymer is first made into sheet by a prior rolling technique known as calendering, and is subsequently passed between rolls at the same time as the substrate material, in a process that resembles roll bonding. The coating thickness can be controlled very closely. Colours can be selectively transferred to the roller and embossed into the coating to produce repetitive patterns. This technique is used to coat a wide variety of materials including fabric, steel and paper. In the latter case, polyethylene is used to provide liquid sealing properties, for use in milk cartons and the like.

plastic spray coating. See *flame spraying of plastic coatings* and *electrostatically sprayed plastic coatings*

plating. Usually referring to any metallic coating produced by electroplating.

platinum plating. An electroplating method utilising a complex sulphato-dinitro platinous acid electrolyte with a platinum content of 5 g/l used in conjunction with platinum or platinised titanium anodes. A plastic or glass lined vessel is required and plating is carried out at 30°C using a current density of 0.5 amp/dm^2. In this way, coatings up to 3 μm thick can be produced after 2.5 hours. The coatings are somewhat harder than pure platinum and are used for plating jewellery, electrical contacts and scientific instruments.

pole figures. A standard method of quantifying the preferred crystallographic orientations of bulk and surface modified materials by measuring the intensity of X-rays diffracted from a surface for a specific diffraction condition, using an X-ray pole figure goniometer. The intensity of X-rays diffracted by a specific family of crystallographic planes of any given coating can be measured over a complete range of angles from the surface normal (90°) to about 5° above the sample surface. An example of a pole figure for a TiN coating, deposited by CVD is given in the citation for *texture.* Although three pole figures are usually required to provide a complete description of crystallographic texture, one pole figure is considered adequate for fibre textured coatings, such as those commonly produced by CVD and PVD.

polymer coatings. See *thermoplastic polymer coatings* and *thermosetting polymer coatings.*

polytetrafluoroethylene coating. See *PTFE coating.*

porosity. Porosity frequently occurs in surface engineered materials, epecially in thermal sprayed coatings (see *thermal shock resistance*). Poor selection of power beam paramaters can also cause porosity to occur in power beam surface alloyed or melted surfaces. Sometimes, porous surfaces are deliberately encouraged, e.g., in the *Nitrotec* process, micropores are developed in the compound layer which serve to retain wax for the purpose of imparting corrosion resistance. Also see *coating porosity* and *laser healing*.

porous chrome plating. See *chromium plating*

powder coating with plastics. See *electrostatic fluidised-bed coating* and *fluidised-bed coating with plastics.*

powder flame spray deposition of plastic coatings. See *flame spraying of plastic coatings*

powder flame spraying. See *thermal spraying* and *thermospraying*

powder metallising. Any thermal spraying method in which a metallic material is deposited.

power beam surface alloying. Any treatment involving localised liquid-phase alloying of (usually) metallic surfaces. A high power density beam (laser or electron) provides the thermal energy required for melting; in the case of laser treatments oxidation protection is provided by a He gas shroud, electron beams are operated in-vacuo. Alloying elements can be injected into the molten pool as a powder (termed "particle injection") or as a gas (e.g., laser nitriding or electron beam gas alloying); alternatively, a solid coating can be pre-placed on the metallic component and subsequently "melted-in". The latter provides better control with respect to the amount of material being alloyed. Such techniques are highly 'line-of-sight' and are most easily applied to external surfaces.

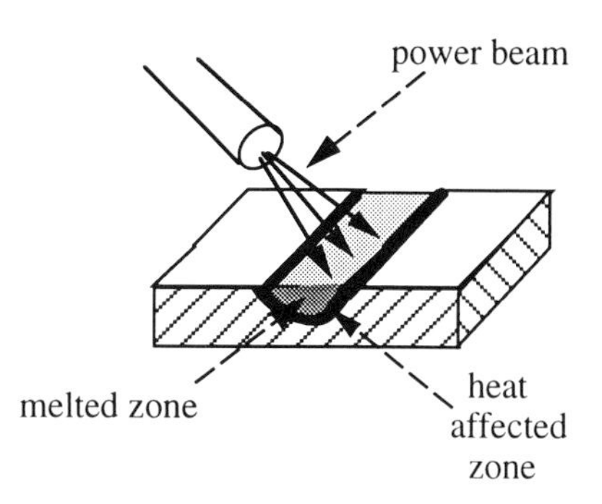

As with all other power beam methods, a treated surface must be progressively built-up by a series of overlapping tracks. Track widths vary according to the type of laser/electron-beam, interaction time and power density, but are typically ~0.5 to 2 mm. Hence, numerous power beam passes are required before a treatment is completed. However, because power beam scan speeds are frequently rapid, economic treatment times per unit component are still possible

power density. Commonly expressed in W/cm^2. This parameter is of great importance in plasma, laser and electron beam methods.

p.r.c. See *periodic reverse current.*

preferred orientation. See *texture.*

pressure nitriding.

> 'Gas nitriding carried out at higher than atmospheric pressure' – IFHT DEFINITION.

The pressure used in this process depends upon the ratio of the charge surface area to the volume of the nitriding vessel. Approximately 50 to 100 g of ammonia/m^2 of charge are used.

primer. The first layer of a multilayer organic paint, its main purpose being to wet the surface and provide adhesion for subsequent coatings and to seal the surface thereby providing corrosion protection.

primer coating. See *primer.*

progressive hardening. See *flame hardening.*

prompt radiation analysis. See *NRA*.

protective atmosphere

> 'An atmosphere which prevents objects from undergoing such reactions as oxidation or decarburisation during heat treatment'
> – IFHT DEFINITION.

protective coating. Any coating that serves to protect a material from degradation through wear, corrosion or oxidation.

PSII. Plasma source ion implantation. *See plasma immersion ion implantation (PI^3)*

PTFE coating. A 'non-stick' or 'release' coating of polytetrafluoroethylene (PTFE) applied to metallic surfaces, commonly by a two-step procedure. Finely divided PTFE, suspended in an appropriate fluid, is 'painted' or sprayed onto a metallic surface previously roughened by grit blasting or similar; this enables reasonable coating adherence through keying-in. The deposit is allowed to dry and the PTFE is subsequently fused by heating to temperatures ~400°C. Various proprietry compositions exist, which vary in adherence and durability. Coatings are typically ~ 20 µm thick. Although these products continue to improve, coating adherence still represents an engineering challenge which frequently limits their application range. Also known as Teflon® coatings. (Teflon is a registered trademark of the DuPont Company.)

pulsed laser. See *pulse mode laser.*

pulse electroplating. See *pulse plating.*

pulse mode laser. Any laser in which the coherent beam of laser radiation is generated as regular intense pulses of short duration (~ 10 to 40 ns). This mode of operation is helpful for laser glazing where high power densities coupled with short interaction times are required. Excimer and Nd-YAG lasers generally operate in this mode. The power delivered in any given pulse far exceeds that which is possible in the continuous wave (CW) mode of operation. For example, consider an excimer laser operating with a pulse duration of 40 ns and delivering 4J of energy per pulse; this equates to 10^8 watts of power per pulse. Conservatively estimating the beam diameter at 0.1 mm, gives a power density of 3.2×10^{11} W/cm^2! Also see *laser.*

pulse plasma. A glow discharge plasma generated by a rectified alternating current power supply. The rectification is achieved in such a way so as to enable variation in the frequency of each DC pulse, pulse shape and pulse duration. Frequencies can be varied from a few Hz through to the radio frequency (MHz) range. There is some considerable controversey regarding the virtues of such devices, in comparison to full wave rectified (FWR) devices, especially for use in plasma nitriding. In the latter case, auxiliary radiant heaters are required since less thermal energy is transferred to the workpiece surface per unit time (via ionic bombardment) than for plasmas driven by FWR power supplies. It is also claimed that pulse plasmas have superior throwing power enabling the penetration of deep holes.

pulse plating. A relatively recent development in the electroplating field (circa 1979) in which a pulse power supply is used to provide a pulsating current. Three advantages are claimed for this approach: (i) the ability to plate at high average cathodic current density; (ii) a notable refinement in the grain size of the electrodeposits; (iii) an improvement in throwing power (particlularly on the micrometer scale).

PVD. See *physical vapour deposition (PVD).*

pyrolytic CVD. Any CVD process that utilises a pyrolitic reaction (or reactions) for the deposition of a material or coating. Probably the most widely applied pyrolysis reaction is that of methane, used in the production of pyrolytic carbon:

$$CH_4 + \textit{heat} \longrightarrow C + 2H_2$$

When carried out at 2200°C at reduced pressure (~10 torr), pyrolytic carbon has a similar structure to graphite in that its basal planes are roughly parallel; this material has a characteristic 'c' lattice cell dimension of 0.344 nm, compared to 0.335 nm for graphite. Pyrolytic carbon deposits can be produced at temperatures as low as 900°C; but these materials are less strongly oriented.

Some compounds that are suitable for pyrolytic CVD are frequently unstable. Diborane (B_2H_6), for example, detonates on contact with moist air. Such properties demand stringent process control protocols. Diborane readily undergoes pyrolysis between 500 and 600°C according to the reaction:

$$B_2H_6 + \textit{heat} \longrightarrow 2B + 3H_2$$

The resulting coatings are poorly crystallised, i.e., they contain significant quantities of amorphous boron. If required they can be fully crystallised to β-boron by vacuum annealing at 1200°C for 30 minutes.

pyrolytic plating. See *pyrolytic CVD.*

Q

quench-hardened layer

> 'Surface layer of an object hardened by quenching from the austenitic condition, generally defined by the effective depth of hardening' – IFHT DEFINITION.

R

radiation cure coating. A plastic coating applied while in the monomer state. On exposure to ultraviolet radiation the coating undergoes a photochemical reaction and becomes polymerised.

radio frequency (RF) bias ion plating. *See radio frequency (RF) ion plating.*

radio frequency (RF) bias sputter deposition. Sputter deposition in which the bias applied to the objects/components is in the radio frequency range (typically 13.56 M. Hz). Regarding the mode of target sputtering there are two possibilities: (i) operation with a DC glow discharge plasma (for conducting targets); (ii) operation with an RF glow discharge plasma (for non-conducting targets). RF bias sputtering is used when the objects being coated are insulators or when the coating being deposited is electrically insulating. However, it is sometimes used to 'etch', i.e., depassivate metal surfaces before proceeding with DC bias sputtering.

radio frequency (RF) glow discharge. Any glow discharge plasma in which the bias potential is cycled at very high frequency (radio frequency) typically within the range of 10 kHz to 1000 MHz, although, one very common frequency is 13.56 MHz. This is one of the permitted radio frequencies set aside for non-radio transmission purposes, to enable use without disturbing radio communications. An important feature of RF plasmas is that charge cannot build-up on an insulating surface in the way that would happen when placed in a DC glow discharge plasma; consequently the latter plasmas quickly extinguish. RF plasmas require tuning or matching to any given processing chamber and the cost of RF power supplies, per unit power, is approximatly double that of DC power supplies. Accordingly, this has restricted in the industrial implementation of RF plasma systems to very specialised sectors, such as thin film device manufacture.

radio frequency (RF) ion plating. Ion plating in which an RF bias is applied to the objects/components in place of the usual negative (DC) bias. It is used either if the coating being deposited is an electrical insulator (e.g., dielectrics like ZrO_2 or Al_2O_3) or when the objects/components are themselves non-conducting (e.g., dielectrics or polymers). Also see *radio frequency (RF) glow discharge.*

radio frequency (RF) magnetron sputtering. Magnetron sputtering that utilises an RF glow discharge plasma to sputter insulating targets such as Si, Si_3N_4, ZrO_2 or Al_2O_3. Also see *radio frequency (RF) bias sputter deposition.*

radio frequency (RF) PAPVD. Any plasma assisted PVD method utilising an RF glow discharge, e.g., see *radio frequency (RF) ion plating* , *radio frequency (RF) sputter deposition* and *radio frequency (RF) bias sputter deposition.*

radio frequency (RF) plasma. See *radio frequency (RF) glow discharge.*

radio frequency (RF) plasma CVD. Plasma assisted CVD in which the frequency of the bias potential, used to creates an RF glow discharge, is within the radio frequency range (typically from 10 kHz to 1000 MHz). RF plasma CVD is used for the deposition of insulators (dielectrics like Al_2O_3) to enable mass transfer by the plasma to the deposition surface without the problem of charging. It has also been found useful for the deposition of metal carbides and nitrides from metal halides (like $TiCl_4$) which are problematic for sustaining DC glow discharges due to their propensity to break down and form arc discharges. Also see see *radio frequency (RF) glow discharge.*

radio frequency (RF) sputter deposition. Sputter deposition that utilises an RF glow discharge plasma to sputter insulating targets such as Si, Si_3N_4, ZrO_2 or Al_2O_3, without magnetron enhancement. Also see *radio frequency (RF) bias sputter deposition.*

rattling. See *tumbling.*

RBS. Rutherford backscattering spectrometry. This method requires a very high energy (0.5-4.0 MeV) particle (H^+, D^+, $^3He^+$,$^4He^+$) beam which is directed normal to the sample surface. The ions collide elastically with the lattice atoms of the sample. Many are back-scattered and subsequently collected by a detector which counts their number and measures their energy (diagram). It is an important feature of RBS that the energy of the back-scattered particles can be used as a basis to calculate the *depth and/or mass* of the sample atoms that caused the scattering. During RBS the amplified detector signal is collected over a period of time by a multi-channel analyser. The 'raw data' is subsequently displayed as yield (proportional to the total number of back-scattered particles) against channel number (proportional to energy). These results are then quantified with well proven computer simulation programs (e.g., see J. C. B. Simpson and L. G. Earwaker, *Vacuum,* 1984, **34**, 899), which through iterative calculation enables the derivation of depth concentration profiles. RBS is particularly suited to determining concentrations of high atomic number elements dispersed in a low atomic number matrix. The maximum useful detection depth (below the surface) is about ~1 μm. Also see *NRA* and *PIXE.*

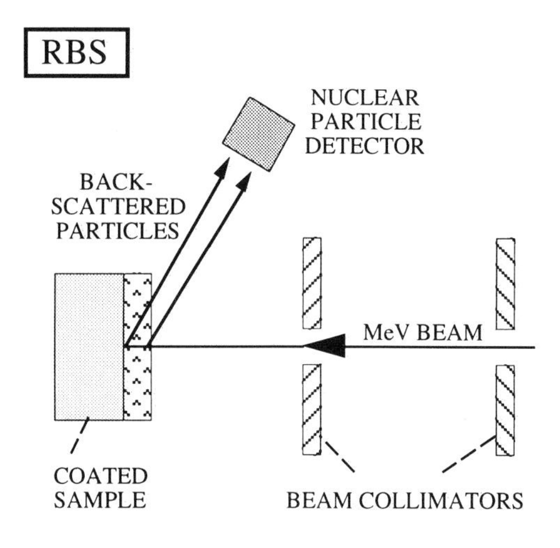

reactive atmosphere. Any atmosphere which reacts with a surface or that combines with other species to form a surface coating or diffusion zone. For example, controlled reactive atmospheres form the basis of gaseous nitriding, gaseous carburising, reactive ion plating and reactive sputter deposition.

reactive dip coating. See *Toyota diffusion (TD) process.*

reactive evaporation. Evaporation coating whereby metal is vaporised into a reactive atmosphere thereby forming a ceramic coating. The substrate may be resistively heated. This process is distinguished from reactive ion plating in that it is carried out in the absence of a glow discharge plasma.

reactive evaporative PVD. See *reactive ion plating (RIP).*

reactive ion plating (RIP). Ion plating (a plasma assisted PVD method) in which a reactive gas is introduced into a glow discharge plasma (surrounding the objects/components) which reacts with the evaporant, forming a compound coating. For example titanium vapour can be reacted with nitrogen to form a titanium nitride coating. One of the more widely exploited forms of plasma assisted PVD. Also see *ion plating* and *evaporative source PVD.*

reactive medium

> 'Solid, liquid, gaseous, or plasma medium used in a thermochemical treatment' - IFHT DEFINITION.

Any medium which reacts with a surface or that combines with other species to form a surface coating or diffusion zone. Controlled reactive media form the basis of many techniques, especially thermochemical diffusion and deposition methods.

reactive PVD processes. See *reactive ion plating (RIP)* and *reactive sputtering*

reactive sputtering. Any sputtering method whereby multicomponent coatings, such as interstitial compounds, can be produced by sputtering into a reactive atmosphere, e.g., transition metals can be sputtered into nitrogen or carbon enriched atmospheres to produce, respectively, transition metal nitrides (diagram) or carbides according to the following generalised equations (where M equates to the metal component):

$$2M + N_2 \longrightarrow 2MN$$
$$M + CH_4 \longrightarrow MC + 2H_2$$

Reactive sputtering is deployed in both magnetron and non-magnetron (diagram) sputtering modes of operation. The former is preferred because of its higher deposition rate. See *magnetron sputtering*. The diagram is based on an earlier version by John Thornton.

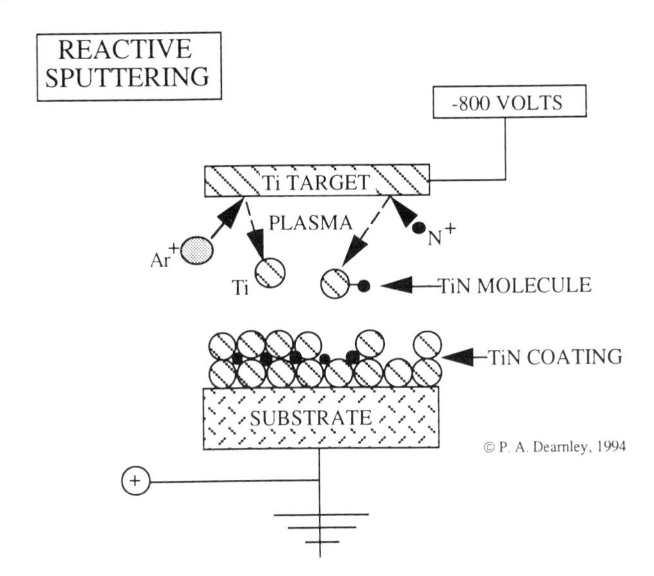

reactive surface

> 'Region of the surface of an object which reacts with the surrounding medium during thermochemical treatment'
> – IFHT DEFINITION

reclamation. The application of (usually) thermal spraying methods for the dimensional restortation of costly engineering components. Large diameter rolls, used in the paper industry, are frequently restored in this way. Other components that are reclaimed include propeller shafts for ships and high precision milling cutter bodies used for metal cutting. Thermal spray techniques are replacing electrodeposition methods for reclamation applications. This is because of the greater diversity of materials, both metallic and ceramic, that can be deposited by the former. Most electrodeposition is confined to depositing metallic coatings, or at best, metal-ceramic composites. Thermal spray equipment can be transported to almost any location. The mobile variant of electroplating is *brush plating*. However, it is similarly limited in its coating material capability; the surface area coverage, per unit time, is also significantly less than that of thermal spray coating techniques. Hence, the expansion of the thermal spray restoration business.

reflectance. The ability of a surface to reflect radiation. Expressed as the ratio of the intensity of the reflected radiation to the intensity of the incident radiation.

reflectivity. See *reflectance.*

residual stress

> 'Stresses that remain in an object when all externally applied forces and temperature differences are absent' – IFHT DEFINITION.

Residual elastic stress (actually strain) within a material; subdivided into macrostress and microstress. Macrostress is residual elastic stress (strain) that occurs in a very large number of grains and is transmitted across adjacent grain boundaries. Microstress is elastic stress (strain) confined to individual grains that is *not* transmitted across adjacent grain boundaries. Both types of stress are best measured using X-ray diffraction methods. Macrostress is determined from the peak shift observed when an object (sample) is rotated through φ (the angle between the plane normal and the surface normal) while fixed in a specific diffraction condition (i.e., at a specific 2θ value); a plot of $\sin^2\varphi$ versus $(d_\varphi - d_0)/d_0$ is obtained, where d_0 is the unstrained d-spacing and d_φ,the strained d-spacing at a surface angular rotation of φ. The slope of this plot is used to calculate the residual macrostress, provided Young's modulus (E) and poissons ratio for the material under investigation is known. Microstress is determined from a numerical analysis of the peak broadening observed at very high angles of 2θ. Other methods of quantifying residual stress also exist; these encompass ultrasound, neutron diffraction, microdrilling and magnetic field techniques.

resistive evaporation source. Used in ion plating type PVD processes in which the metal or alloy being evaporated is in contact with a resistively heated crucible, filament or 'boat'. This process is necessarily restricted to those metals or alloys which evaporate at a temperature below the melting point of the crucible, filament or boat; it is used in Ivadising for the evaporation of aluminium since it offers a cheaper alternative to electron beam or arc source evaporation technologies. Care in selection of the correct crucible material is required, in order to avoid detrimental reaction, especially those arising whilst the evaporant is in the molten state.

retained austenite.

'Austenite remaining at ambient temperature after a quench-hardening treatment' – IFHT DEFINITION

Moderate quantities of retained austenite (α) are unavoidable in martensitic (β) carburised cases. When retained austenite exceeds ~ 50 vol% there is a marked reduction in case hardness and fatigue strength (rolling contact and rotation-bending). The usual cause of excessive retained austenite is a too much case carbon (> 0.9wt-%) caused by a too high carbon potential. Retained austenite is always most noticable at component corners, where, during carburising, carbon is being supplied via two faces. In the worst examples, massive, blocky iron carbides can from. The quantification of retained austenite is best achieved by using an X-ray diffraction procedure. Commonly the ratio of the intensities of the $\{200\}_{\alpha'}$ to that of the $\{111\}_{\gamma}$ are used for this purpose.

reverse current cleaning. See *electropolishing.*

RF plasma CVD. See *radio frequency (RF) plasma CVD.*

rhodium plating. An electroplating process in which rhodium is deposited from solutions containing 2 g/l Rh and 20 ml/l concentrated sulphuric acid. Electroplating baths are operated at 35°C with a current density of 0.5 to 1 amp/dm^2. Rhodium has an inherently high chemical stability being resistant to all acids and other corrosive materials at ambient temperature. Rhodium plating is used in the electrical and electronics industries for the protection of electrical switch contacts used in oxidising or corrosive environments; it also finds use for the protection of silverware (to prevent tarnishing) and jewellry. Rhodium electroplates have a Vickers hardness of ~800 kg/mm^2, are white in colour and are highly reflective.

RIP. See *reactive ion plating.*

rod flame spraying. See *combustion wire gun spraying.*

roller coating with plastics. See *plastic roller coating.*

roll hardening. A method of hardening surfaces through localised cold working. In the case of cylinderical objects, the surface is pressed between three rolls, which are rotated over the surface, while applying sufficient pressure to cause plastic deformation. The cold worked surface remains in a state of residual compressive stress, imparting a marked increase in fatigue strength. This technique is applied to high strength steel, nickel and titanium alloy bolts and fastners (see *micrograph*). High strength bolts, used for securing engine heads, are often strengthened below the bolt head using roll hardening; it is very effective in preventing dangerous fatigue failures. For a practical review of this technology, see: G. Turlach, *Surf. Eng.*, 1985, **1**, 17-22.

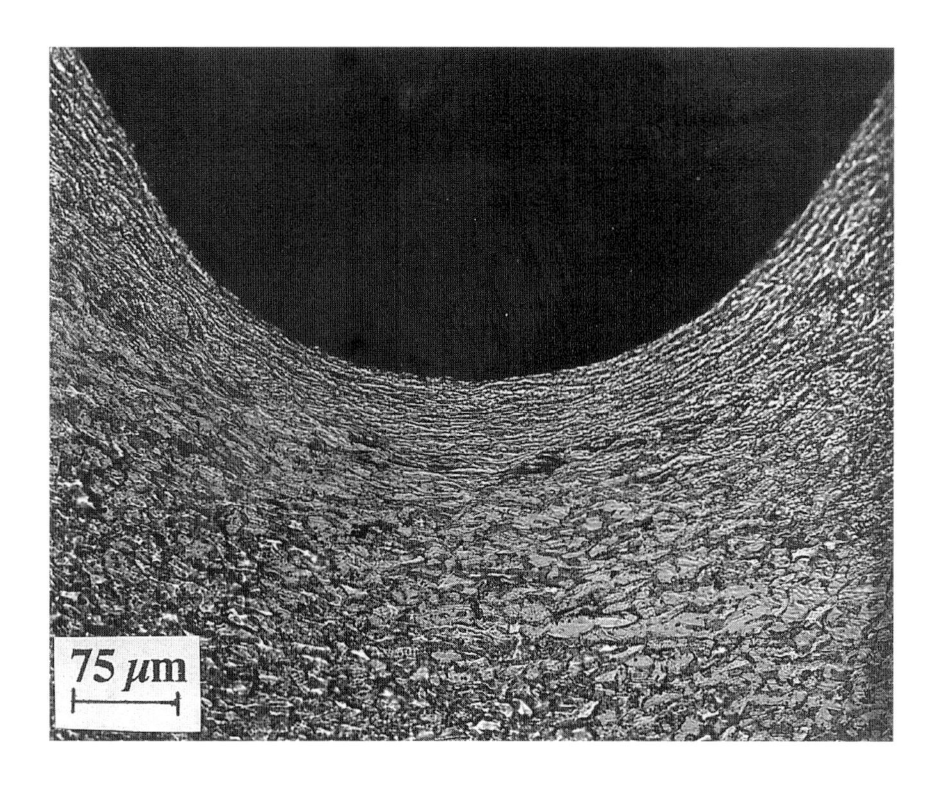

roll hardening. *Microsection through the thread root of a roll hardened Ti-6Al-4V fastener. Note the intensive cold working. Etched in 1.5 % HF.*

rolling contact fatigue. An effect that takes place by the conjoint action of rolling contact and traction stresses as typified in the operation of gears, i.e., where there is continous contact between two surfaces moving at slightly (~10%) dissimilar surface speeds. Nitriding and carburising are probably the two most important surface engineering methods that markedly improve the rolling contact fatigue endurance of steels. The diagram (based on data from M. Weck and K. Schlotermann, Metallurgia, 1984, No 8, 328-332) shows the effectiveness of plasma nitriding in raising the fatigue strength and endurance of gear teeth made from prior hardened and tempered low alloy steels (designated DIN 31CrMoV9V and 16MnCr5N); note: the nitrided case depth is given in parentheses. The plain carbon steel (DIN CK45), containing no nitride forming elements, has a fatigue strength and endurance only slightly better than when in the untreated condition; essentially, this material was in the nitrocarburised state and had an outer surface layer of ε-carbide, several micrometers thick.

ROLLING CONTACT FATIGUE

GEAR TOOTH ROOT FATIGUE ENDURANCE FOR PLASMA NITRIDED AND NITROCARBURISED STEELS.

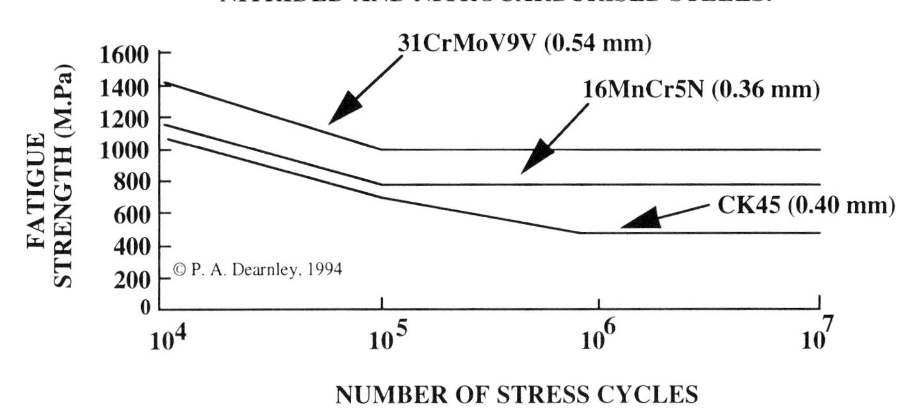

rolling contact wear. In the case of untreated steel surfaces, rolling contact fatigue stresses result in the formation of localised surface failures, which result in cavities ~ 20 to 100 µm across, being formed. For treated (carburised or nitrided) or untreated steels, superficial surface plastic deformation can also take place. In the absence of an oil based lubricant, mild or severe oxidational wear also contributes to the overall wear.See Y. Sun, P. A. Dearnley and T. Bell, *Proc.Plasma Surface Engineering,* **2**, 927-934, 1989, Oberursel, DGM.

rose gilding. See *colour golds.*

roughness. See *surface roughness.*

ruthenium plating. An electroplating process in which ruthenium is deposited from electrolytes based on ruthenium sulphamate or ruthenium nitrosylsulphomate, although it is claimed that better deposits result when the electrolyte is based on the anionic complex $(H_2O. Cl_4 . Ru . N . Ru . Cl_4 . OH_2)^{3-}$. Deposits up to 5 µm thick are usually produced. Ruthenium plating has potential for replacing gold and rhodium plating, especially for electrical 'sealed-reed' relay switch applications.

Rutherford backscattering spectrometry. See *RBS.*

S

S-phase. A term first coined by Ichii and others (Faculty of Engineering, Kansai University, Osaka, Japan) to describe the shallow (~3 μm) surface zone produced in the absence of CrN after nitriding austenitic stainless steels at or below 500°C. Also referred to as 'expanded austenite'. The precise nature of S-phase has yet to be conclusively elucidated, but it is characterised by exceptional hardness (~1300 kg/mm^2), high residual stress and outstanding corrosion resistance. It displays only three or four broadened diffraction peaks when exposed to monochromatic X-rays; the peak positions being consistent with a tetragonal unit cell. The phase is meta-stable; it is not observed when nitriding at ≥550°C. Instead, the chromium is partitioned out of solution as CrN. In plasma nitriding, relatively high power densities are required to induce the formation of S-phase. Also there has been some practical experience with lack of uniformity of treatment, especially when treating large numbers of components. The diagram shows potentiodynamic sweeps (at 0.5 mV per second) for AISI 316 austenitic stainless steel plasma nitrided in cracked NH_3 for various temperatures and times, e.g., 800/10 indicates 800°C for 10 hours. All tests were carried out in 3% NaCl solution at ambient temperature relative to a Standard Calomel Electrode. The data shows that all samples containing S-phase (500/10, 500/36 and 450/36) have superior corrosion resistance to non-nitrided AISI 316. Samples nitrided at 600°C and above (where CrN predominantly forms) exhibit inferior corrosion resistance. Data from P. A. Dearnley, A. Namvar, G. G. A. Hibberd and T. Bell, *Proc.Plasma Surface Engineering,* **1**, 219-226, 1989, Oberursel, DGM. S-phase/expanded austenite or its analogues can be produced by plasma immersion ion implantation (PI^3) of nitrogen into austenitic stainless steel at 350 or 450°C (see M. Samandi, B. A. Shedden, D. I. Smith, G. A. Collins, R. Hutchings and J. Tendys, *Surface and Coatings Technology*, 1993, **59**, 261-266).An interesting alternative is the possibility of producing S-phase type coatings via reactive magnetron sputter deposition. The earliest work on this appears to have been done by Bourjot and co-workers (see A. Bourjot, M. Foos and C. Frantz, *Surface and Coatings Technology,* 1990, **43/44**, 533-542).The advantage here is that relatively thick layers of this material can be produced (~5-20μm) in a comparatively short time (~1-2 hours), unlike those formed via plasma nitriding or PI^3 which are necessarily limited in thickness by the diffusion kinetics pertaining at 350 to 500°C.

S-PHASE

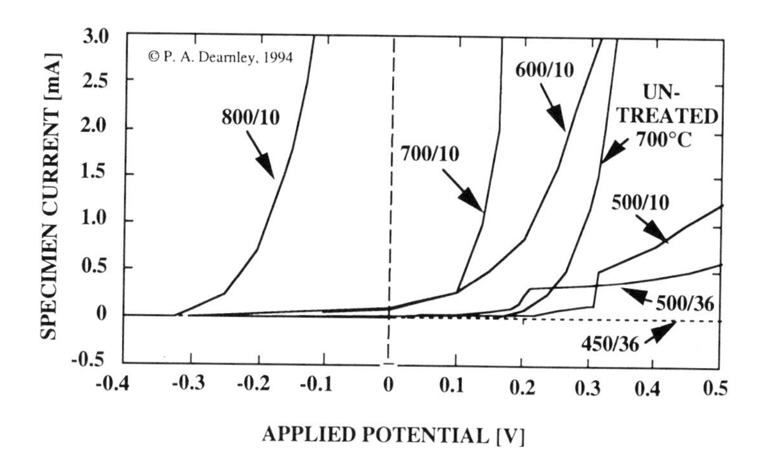

sal-ammoniac tinning. A variant of single-pot tinning. Fluxed steel objects are dipped into a bath of molten tin covered by powdered NH_4Cl, which serves to prevent oxidation.

salt bath aluminising. Carried out electrolytically or electrolessly. In the electrolytic method, fused salt mixture containing, typically, NaCl, KCl, and Na_3AlF_6, at a current density of 0.5-4.5 A/dm^2 are employed. The electroless method entails immersion in a fused eutectic of 2:1 KCl_3- LiCl, and AlF_3 at temperatures ~680-950°C for 4-5 hours. The resulting layer is claimed to exceed 300 µm. A rarely practised method; hot dip aluminising is more popular.

salt bath boriding

> 'Boriding carried out in a borax-based medium' – IFHT DEFINITION.

There are two principal types of liquid phase boriding. Electrolytic and electroless. Both are best carried out at temperatures above 900°C such that salt bath viscosity is low enough to obviate compositional 'dead zones'. Such baths operate fairly close to the melting point of borax (741°C); the principal bath constituent. Both techniques are rarely practised outside Eastern Europe.

Electrolytic boriding is the oldest commercial boriding technique which was devised in the former USSR in the 1930s. The original method is reliant on the electrolysis of fused borax ($Na_2B_4O_7$) but the mass transfer of boron takes place in two distinct steps. The component is attached to the cathode where sodium ions are neutralised to form liquid sodium. The first step in the boron transfer mechanisms is at at the graphitic anode where the tetraborate ions are neutralized to form boric acid and nascent oxygen:

$$B_4O_7^{2-} \longrightarrow B_4O_7 + 2e \longrightarrow 2\,B_2O_3 + O$$

The nascent oxygen reacts with the anode and produces CO. The second step involves the diffusional migration of the neutral boric acid molecules towards the cathode where they become reduced by the liquid sodium to form boron:

$$6Na + B_2O_3 \longrightarrow 3Na_2O + 2B$$

The boron subsequently diffuses into the ferrous components. Current densities should not exceed 0.1 A/cm^2 otherwise non-uniformity of treatment results. A more recent electrolytic salt bath method has been devised that enables lower treatment temperatures to be used (See *metalliding*).

Electroless boriding relies on the reduction of molten borax by SiC and/or B_4C. However, it has been found preferable to use ~30wt-% B_4C as a reductant since bath viscosities are lower and throwing power is improved. Components are simply immersed in the fused salt mixture and their surfaces quickly become borided. Work in Japan has shown that boron potential can be enhanced by replacing up to 20wt-% B_4C with ferroaluminium. More recently salt baths comprising 75 wt-% KBF_4 -25wt-% KF have been used to boride nickel alloys at temperatures below 670°C. Also see the review: P. A. Dearnley and T. Bell, *Surface Engineering,* 1985, **1**, (3), 203-217.

salt bath calorising. See *salt bath aluminising.*

salt bath carbonitriding. See comments on *carbonitriding* and *salt bath carburising.*

salt bath carburising

'Liquid carburising carried out in a molten salt mixture'
– IFHT DEFINITION.

Mainly carried out as an electroless immersion technique for small parts that in general require a case depth of less than 0.5 mm. Prior to immersion, parts are pre-heated in other salt baths held at ≈150 - 400°C. Conventional salt bath carburising is carried out in fused sodium cyanide (NaCN) based salts. For operation within the range 850–900°C a typical salt bath comprisies (by weight) 10-50% NaCN, 0-25% potassium chloride (KCl), 20-40% sodium chloride (NaCl), 30% sodium carbonate (Na_2CO_3) and 1% sodium cyanate (NaCNO). Several reactions are important. At the fused-salt/atmosphere interface the bath undergoes the ageing reaction:

$$2NaCN + O_2 \longrightarrow 2\,NaCNO \quad \text{——(1)}$$

This compound decomposes to liberate carbon monoxide (CO) and nascent nitrogen (N) at the fused-salt/component interface:

$$4\ NaCNO \longrightarrow 2\ NaCN + Na_2CO_3 + CO + 2N \quad \text{——(2)}$$

The CO is reduced by the steel workpiece and the C is taken into solution:

$$2CO \longrightarrow C_{(Fe)} + CO_2 \quad \text{——(3)}$$

From equation 2 it should be noted that nitrogen is also available for solution. The amount of carbon and nitrogen dissolved by the steel charge is largely determined by temperature and the amount of NaCN. High carburising temperatures (>850°C) and NaCN contents above 50%, lead to a maximisation of carbon up-take at the expense of nitrogen, i.e., the nitrogen up-take decreases. This is in compliance with the decreasing nitrogen solid solubility with ascending temperature, for this particular temperature range. For example, after salt bath carburising a plain carbon steel at 800°C in 50% NaCN for 2 hrs the surface concentrations of carbon and nitrogen have been observed to be 0.73 and 0.55 wt-% respectively. Using the same NaCN content and duration of treatment at 900°C the surface concentrations of carbon and nitrogen were determined to be 0.91 and 0.20 wt-% respectively.

Following carburising, the charge is removed and directly quenced into oil. Salt bath carburising although attractive in some ways (e.g., its rapidity of heating) has the major disadvantage of presenting major waste disposal problems. Consequently, non-cyanide based salt bath compositions have been developed comprising halide compounds with graphitic additives. The process economics of these new treatments have yet to be revealed; it is unlikely that they will compete with the larger scale vacuum, gaseous and plasma carburising methods, but are a welcome replacement for the hazardous sodium cyanide technique.

salt bath chromising. Carried out in a molten salt mixture containing appropriate chromium compounds at 900 to 1200°C for ~ 6 hours. Rarely practised outside Eastern Europe.

salt bath nitriding/nitrocarburising. The distinction between salt bath nitriding and salt bath nitrocarburising only resides in the type of substrate treated and not in the composition of the salt bath used, which is identical. Salt bath nitriding applies to low alloy (1-3%Cr, 0.5% Mo) steels which develop a characteristic high strength diffusion zone, whereas, salt bath nitrocarburising applies to plain low-medium carbon steels, which develop a less strong diffusion zone. In both cases a wear resistant compound layer, enriched in ε-$Fe_{2\text{-}3}N$ is produced on the surface.

In the established method of salt bath nitriding/nitrocarburising the fused salt comprises 60-70 wt-% sodium cyanide (NaCN) and 30-40 wt-% potassium cyanide (KCN). In addition there are minor additions (a few per cent) of sodium carbonate (Na_2CO_3) and sodium cyante (NaCNO). By ageing the bath at 575°C for 12 hours the cyanate content can be raised to the working level of about 45 wt-% via the oxidation reaction:

$$2NaCN + O_2 \longrightarrow 2\ NaCNO$$

The cyanate then decomposes at the component-fused salt interface via the catalytic reaction:

$$4NaCNO \longrightarrow 2NaCN + Na_2CO_3 + CO + 2N$$

Salt bath nitriding/nitrocarburising is carried out at the upper end of the nitriding temperature range (~550 to 570°C) for relatively short times (<2 hours), since the nitrogen potential is relatively high (compared to gaseous and plasma techniques). The active carbon and nitrogen in the process serve to stabilise the ε-$Fe_{2-3}N$ phase in the external compound (or white) layer in preference to the γ'-Fe_4N; the latter is found in minor amounts. Refinements of the process have been introduced, of particular note are *Sulfinuz* and the *Tufftride Process*.

salt bath siliconising. Carried out electrolytically or electrolessly. The source of silicon is an alkaline metal silicate, ferrosilicon, crystalline silicon or silicon carbide. Processing temperatures are ~900–1100°C for 0.5–10 hours. Rarely practised outside Eastern Europe.Also see *siliconising*.

sand blasting. Grit blasting with sand particles.

sandwich coating. See *composite coating* and *modulated coating*.

satin surface finish. See *matt surface finish.*

saturation coverage. Complete surface coverage by any given treatment.

scale resistance. See *oxidation resistance.*

scanning auger microscopy (SAM). A development in auger electron spectroscopy (AES) that enables secondary electron imaging and elemental auger electron mapping of selected areas. Some instruments have limited spatial resolution making satisfactory secondary imaging beyond approximately x10,000 difficult. The technique is very useful for mapping the distribution of 'light' elements like carbon, nitrogen and boron.

scanning electron microscopy (SEM). A standard microscopy method with exceptional depth of field. Modern instruments now deploy field emission electron sources which provide excellent illumination. Especially useful for the examination of worn, corroded or fractured surfaces.

scanning induction hardening. Induction hardening in which an object is traversed under an induction coil and is immediately water quenched. This is a standard procedure for the hardening of slide-ways and other machine parts.

scanning tunneling microscopy. A relatively new technique enabling the resolution of lattice points on a sample surface. The method may prove useful in the elucidation of wear and corrosion mechanisms.

scratch adhesion test. A test where a diamond stylus is dragged over a surface. The load on the stylus is progressively increased until the coating fails. Some experimental set-ups enable the recording of acoustic emissions which aid interpretation of the recorded forces; a large 'jump' in acoustic emission often accompanies coating failure. Hence, the force required to cause coating failure is detected. Also see *adhesive (or adhesion) strength tests.*

scuffing. See *seizure.*

sealing. A process where surface connected porosity is closed by mechanical burnishing or rolling (in the case of metallic coatings), or sealed by infiltration with wax, vinyl copolymers or other organic substances (in the case of metallic and ceramic coatings). Such treatments are frequently applied to thermal spray coatings. However, infiltration is a critical component in the *Nitrotec* thermochemical diffusion treatment. Wax infiltration improves corrosion performance. Also see *laser healing.*

seizure. When two formerly sliding surfaces bond or weld together. In the case of metal on metal seizure, a metallurgical bond is made across the contacting interface. Seizure can take place as a result of very high contact loads and/or when lubrication breaks down. Some metals, like titanium and austenitic stainless steel show a greater tendancy to seize (for a given loading situation) than other materials. Seizure may take place intermitently and culminates in localised fracture or tearing. This latter phenonemon is sometimes termed galling or, less appropriately, scuffing. When on a small scale, the resulting wear is sometimes called adhesive wear. In metal cutting a special situation arises where the metal chip or swarf is seized to the tool rake face over most of the contact area. The chip does not slide over the tool surface but is intensely sheared within a narrow shear zone. This results in the creation of sufficient heat to commonly raise the surface temperature of cutting tools to well over 900°C.

seizure resistance. The ability of two surfaces, undergoing sliding or rolling contact, to avoid seizure.

selective carburising. See *localised carburising.*

selective plating. An electroplating technique used to achieve a partial coverage of electroplating. Areas not requiring coating are "stopped-off" with non-conducting lacquer or wax.

self-fluxing overlay coatings. See *fusion hard facing alloys.*

self-tumbling. Tumbling without an added abrasive medium.

semitransparency. A quality of thin vacuum deposited coatings which partly transmit and partly reflect light.

service life. The working life-time of a surface engineered object.

SES. See *spark emission spectrometry*

shear strength. See coating interfacial shear strength.

shell hardening.

'Quench-hardening treatment in which austenitising is restricted to the surface layer of the object' – IFHT DEFINITION.
See *contour hardening.*

sherardising.

'Diffusion metallising with zinc' – IFHT DEFINITION.

Thermochemical diffusion treatment involving the enrichment of a metallic surface with zinc; carried out at temperatures ranging from 500 to 800°C for 2 to 4 hours. The main use of sherardising is to increase the corrosion resistance of components constructed from ferrous alloys, especially those exposed to rain and sea water. The diffusion zone comprises various Fe–Zn intermetallic compounds, depending upon the exact thermal cycle deployed, although, γ-Fe_5Zn_{21} is frequently the dominant phase produced by *pack sherardising* and *gaseous sherardising.* No significant distinction can be made between what might be called liquid phase sherardising and what is actually termed *hot dip galvanising.* Processing temperatures for the latter method are somewhat lower than for *pack sherardising* and *gaseous sherardising.* Consequently, the diffusion layer differs in constitution, comprising several Fe-Zn intermetallic compounds (See *hot dip galvanising*), not usually found in gaseous or pack sherardised surfaces.

shielded arc evaporator. A technique for trapping macroparticles emitted from arc source evaporators. An electrically biased shield is placed between the arc source and the substrates which serves to collect the emitted macroparticles. Also see *filtered arc evaporator.*

shock hardening. A laser surface hardening method that uses very high power densities ~10^8 to 10^9 W/cm^2 , with very short interaction times $\leq 10^{-7}$ seconds, to vapourise a sacrificial overlay coating, and thereby impart shock-waves of such intensity to cause work hardening of the material beneath. Rarely practised.

shorterising. Another term for *flame hardening.*

shot blasting. A technique of dry blast cleaning in which metallic or ceramic particles of a generally spherical shape are used. In addition to the cleaning effect, the shot blasting usually leads to a substantial work hardening of the surface.Also contrast with *grit blasting.*

shot peening. A mechanical surface engineering treatment in which the surface of a metal object is exposed to the action of a stream of hard metallic shot under controlled conditions. This treatment results in increasing the hardness of the surface layer by cold working and inducing residual compressive stresses. The main purpose of shot peening is to increase fatigue strength. Great skill is required in the use of this technique. If the peening intensity is too great, surface cracks result; these can have detrimental consequences. Also see *Almen Number.*

shrouded arc evaporator. See *shielded arc evaporator*

shrouded plasma spraying. A means of limiting oxidation during plasma spraying which offers a cheaper alternative to vacuum plasma spraying. In the preferred design, nitrogen is preheated to ~400°C, and passed in high quantities (~ 50 cu. m/hr) into the space immediately adjacent the plasma 'flame'. It also occupies the spray-substrate interface providing an inert shroud for the solidifying molten spray. Cobalt alloy spray deposits have had their oxide content reduced from 16 (obtained for APS) to 2% by this technique. Variable results have been reported for the deposition of MCrAlY coatings. Like most innovations, however, best results only come when the user learns to optimise the process to his own needs.

siliconising. Less commonly termed Ihrigising. Thermochemical diffusion treatment involving the surface enrichment of a material with silicon. Siliconising is mainly applied to ferrous alloys and refractory metals, its aim being to increase heat (oxidation) resistance, although hardening and therefore wear resistance, is also likely to be improved. Siliconising techniques include *pack siliconising, salt bath siliconising, gaseous siliconising* and *plasma siliconising.*

silver electroplating. See *silver plating.*

silver plating. An electroplating process in which silver is deposited from solutions typically containing 19 g/l silver cyanide, 15 g/l potassium cyanide and 25 g/l potassium carbonate. Silver electroplating is applied for decorative purposes (e.g. for tableware and decorative household articles). It is also used in the electronics industry for electrical contact purposes.

SIMS. Secondary ion mass spectrometry. An incident ion beam of neutral or reactive character is directed at a solid surface causing the removal of surface ions (secondary ions) through sputtering. Although early SIMS instruments used low ion energies, <10keV, it is now more common to use higher energies, ~15 keV. Argon is a principal neutral ion beam, while oxygen and caesium are frequent reactive ion beams; secondary ions of either negative or positive polarity can be produced and SIMS instruments can be configured to detect either. The secondary ions are collected by a mass spectrometer and quantified. Under oxygen ion bombardment the positive secondary ion yield of electropositive elements, like Mg, V, Ti and Cr are at least three orders of magnitude greater than the positive secondary ion yields of electronegative elements like C, S and O. Conversely when caesium ions are used and negative ions are detected, the reverse situation applies, i.e., the secondary ion yields of the electronegative elements are much higher than for the electropositive elements. Regardless of polarity, the secondary-ion intensity I_{A^*} of a given element A, is related to its atomic concentration $[A]$, via:

$$I_{A*} = K.P_{A*}.[A]$$

where K is a constant related to the ion beam density and P_{A*} is the practical secondary ion yield of element A. In practice K is difficult to determine. However, if only the ratio of two elemental concentrations is sought (sputtered under similar conditions), then a knowledge of K is no longer required, since

$$[A]/[B] = I_{A*}.P_{B*}/I_{B*}.P_{A*}$$

Hence, relative quantities are easily determined. Two classes of SIMS are often used; static SIMS and dynamic SIMS. The former entails analysis of the immediate surface; dynamic SIMS involves depth profiling, whereby a crater is progressively sputtered into the sample surface over a period of many minutes. Nowadays, SIMS instrumentation also enables secondary ion mapping of slected areas, greatly improving the versatility of the method. An important advantage of SIMS, compared to AES, is that the surface under investigation need *not* be an electrical conductor.

single-point turning. A metal cutting technique using a screw cutting lathe (turning), whereby only one metal cutting tool (with one major cutting edge) is engaged in cutting at any particular time.

single-pot tinning. Prior to dipping, iron and steel objects are fluxed by: (i) immersion in an aqueous solution of zinc chloride and hydrochloric acid or; (ii) passing the objects through a molten fused layer of, for example, 78 wt-% zinc chloride: 22 wt-% sodium chloride, that floats on the molten tin. The objects remain immersed in the molten tin, held at a temperature between 280 and 325°C, for sufficient time for iron-tin intermetallics to form, which provide a sound foundation on which tin can bond. Single-pot tinning is used as a preparation step for soldering or as a finish coating if surface quality is unimportant. The objects may require de-fluxing after removal from the bath by appropriate washing routines. Also see *two-pot tinning* and *wipe tinning.*

SIP. See *sputter ion plating.*

sliding wear resistance. The wear resistance of two component surfaces sliding over each other under lubricated, wet or dry conditions.

slurry erosion. See *erosion.*

soft spots.

> 'Localised small regions of lower hardness on the surface of a quench-hardened object' – IFHT DEFINITION.

solder plating. See *tin-lead plating.*

solid carburising. See *pack carburising.*

solvent cleaning. Cleaning performed, as a rule, by immersing and soaking an object in organic solvents comprising, aliphatic petroleums, chlorinated hydrocarbons, or mixtures of these two sorts of solvents, often with ultrasonic agitation. For objects which are too large to be immersed, the solvent maybe applied by spraying or wiping. Caution must be observed in the selection and handling of organic solvents due to their carcinogenic and/or explosive properties.

spalling. Breaking away of surface fragments of material during wear. Often invoked to mean flaking-off of a coating. Also termed exfoliation.

spark emission spectrometry. An electric arc (spark) is created between an electrode and a mechanically ground sample surface. The emitted light (optical emission) is collimated and passed through a spectrometer. Each element produces a characteristic optical spectrum the intensity of which is proportional to the quantity of a given element in the sample. The machine must be calibrated against known primary standards. It is the main method used for the routine analysis of steels and can be used for concentration profiling, e.g., of carburised steels. The procedure here involves grinding away known amounts of sample between sequential analyses. Because the increments are relatively coarse, it is less satisfactory than GDOES, WDS or SIMS.

speculum plating. A hard silvery-white electrodeposit having a composition of approximately 42wt-% tin and 58wt-% copper. The coating is an intermetallic of formula Cu_3Sn with a Vickers hardness ~500 kg/mm^2. It is applied to surfaces that are normally prior electroplated with bright nickel. It is often used as a 'gold extender', i.e., 4 μm of speculum plus 1 μm of gold can serve with eqivalence to 5 μm of electroplated gold. Also see *bronze plating*.

spin hardening.

> 'Quench-hardening treatment involving spin heating of the object, followed by immediate cooling' – IFHT DEFINITION.
>
> See *flame hardening.*

spray and fuse process. See description given under *fusion hard facing alloys*

spray coating with plastics. See *flame spraying of plastic coatings* and *electrostatically sprayed plastic coatings*

spray deposit. A coating produced by any *thermal spraying* process.

sputter deposition. A PVD process in which the atomisation of coating material is achieved by sputtering, i.e. by bombarding the solid source (target) with ions and neutrals having high kinetic energy, so that the material is removed from the surface of the source by a momentum transfer process. It is an inherently cooler process than evaporative PVD. Furthermore, the avoidance of a molten pool facilitates the operation of the target in any orientation. Without magnetron enhancement, deposition rates are typically lower than in evaporative PVD (see *magnetron sputtering*). Sputtering enables alloys and compounds to be readily deposited without deviation in composition. Also see *bias sputter deposition* and *unbalanced magnetron.*

sputter ion plating. A term used by some workers to mean sputter deposition or bias sputter deposition. The term does not refer to evaporative source ion plating methods. Although in use for some considerable time, it would be best to refrain from using it, since it undoubtedly causes confusion to some.

standard electrode potential. See *electrode potential.*

steam blueing. See *steam treatment.*

steam tempering. The same as *steam treatment.*

steam treatment.

> 'Thermochemical treatment involving the formation of a blue-black oxide on ferrous objects by holding them in superheated steam'
> – IFHT DEFINITION.

Steam treatment is claimed to impart wear resistance and has been applied to high speed steels as a finishing treatment, although it results in inferior performance compared to PVD – TiN coated drills. It is also sometimes applied to highly polished steel to provide a decorative blue-black finish. Also see *blackening.*

steered arc evaporator. Similar to a conventional arc source evaporator except that an external magnetic field is used to "steer" the arc in regular circular paths, assuring a more efficient target utilisation. Also see *arc source PVD.*

stopping-off. Also called masking. A procedure for preventing surface treatment on a particular area of a component. For the prevention of plasma nitriding or plasma carburising (of steel) a physical mask is usually used, e.g, steel plugs serve to cover holes, while in the case of gaseous or vacuum carburising, copper 'paste' is applied. Electroplating masking procedures are detailed under *selective plating.*

strain (or stress) peening. A treatment in which the object is mechanically strained in tension during shot peening. This enables the creation of intensive residual compressive stresses approaching the yield strength of the material once the applied stress is removed.

stress relieving.

> 'Annealing process involving holding at a high enough temperature and cooling slowly to reduce the residual stresses without significantly affecting the structure' – IFHT DEFINITION.

stylus adhesion test. See *scratch adhesion test.*

sub-layer. One of the layers of a multi-layered coating residing below the external (or outermost) coating.

substrate. The original component or testpiece material, which subsequently becomes modified by any given surface engineering method, such as an appropriate diffusion, deposition or energy beam technique.

Sulfinuz. A variant of salt bath nitriding/nitrocarburising (circa 1947) in which sodium sulphite (Na_2SO_3) is added to the usual sodium cyanide/cyanate composition. Although much initial confusion arose with respect to the role of the sulphur based compound, it is now regarded as mainly serving to activate the conversion of sodium cyanide (NaCN) to sodium cyanate (NaCNO), via the reaction:

$$Na_2SO_3 + 3NaCN \longrightarrow Na_2S + 3NaCNO$$

Sodium cyanate is the active constituent in *salt bath nitriding*.

sulphurising. A process similar to Sulfinuz.

sulphocyaniding. A process similar to Sulfinuz.

sulphidising. A process similar to Sulfinuz.

superlattice coating. Using magnetron sputter deposition technology it is now possible to synthesise modulated coatings based on the TiN-NbN and TiN-VN systems. The coatings comprise overlapping layers of, for example, TiN and NbN, where each layer is of equal thickness and whose composition gradually changes in a near sinusoidal manner in the direction of coating growth. High resolution transmission electron microscopy (see nanographs) has shown that there is no definite interphase-interface. Instead there are significant regions of solid solution nitrides of the (Ti,Nb)N kind, on either side of the pure single metal nitride compositions, across which there is no disturbance in the atomic packing sequence. The compositional "wavelength" can be adjusted, through appropriate process control, to be as little as 8 nm, realising Vickers microhardness values in excess of 5000 kg/mm^2 (see nanographs). The term superlattice, although widely used, is nonetheless somewhat inaccurate; they should be termed modulated coatings. Also see *modulated coating*.

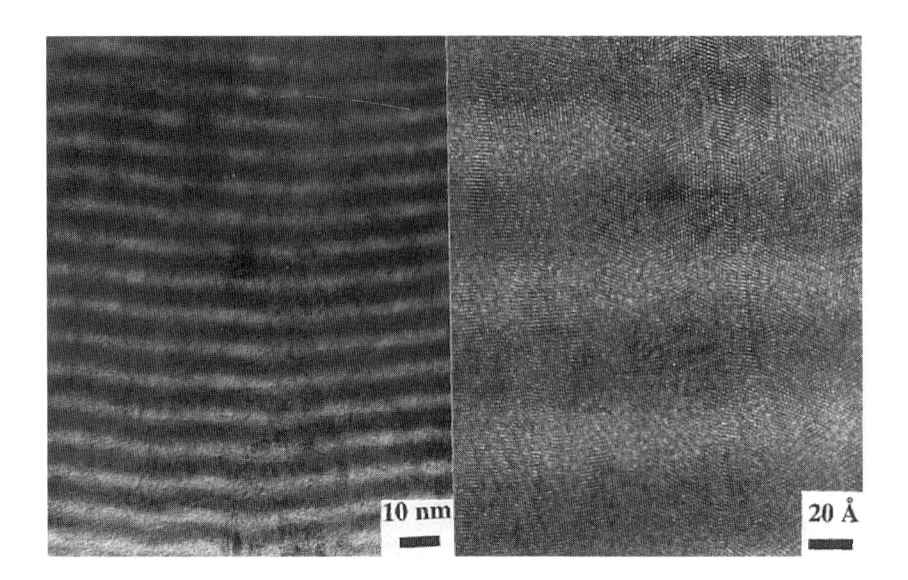

superlattice coating. *High resolution electron microscope images of a 'superlattice coating' formed by the reactive sputter deposition of TiN and NbN using closed field unbalanced magnetrons. The compositional wavelength is 8 nm. Such coatings have outstanding properties; their Vickers microhardness has been determined to be in excess of 5000 kg/mm^2. Courtesy of W. D. Sproul, BIRL, Northwestern University. Used with permission.*

supported glow discharge plasma. A technique whereby additional electrons are supplied to a DC glow discharge plasma (by means of a heated tungsten filament, an electron beam or a closed magnetic field) to enable a relatively high ion density to be achieved for a given working pressure. Ions are generated by electron-neutral atom (molecule) interactions. Supported (or enhanced) glow discharges are used in triode ion plating, activated reactive evaporation, triode sputtering and magnetron sputtering. One benefit is to raise the current density at the substrate surface, improving coating adherence and *throwing power* of the process. It also extends the stability range of DC glow discharge plasmas into the 10^{-3} to 10^{-4} torr pressure range.

surface alloying.

> 'Heat treatment in which the chemical composition of an object is intentionally changed by the diffusion of one or more elements into the surface' – IFHT DEFINITION.

The physical principle behind many surface engineering techniques. This is achieved either through liquid state or solid state alloying. The most common form of liquid state alloying is *power beam surface alloying*. There are many treatments that utilise solid state alloying;

these can involve: (i) interstitial diffusion, like, *nitriding* or *carburising* of steels or; (ii) substitutional diffusion, in the case of *chromising* or *aluminising* steels. Surface alloying permits a variety of surface properties to be developed enabling resistance to corrosion, wear or rolling contact fatigue.

surface annealing. Any thermal treatment that results in a shallow surface zone being recrystallised or stress relieved. This can be achieved using high frequency induction or localised power beam methods.

surface coating. See *coating.*

surface coverage. A qualitative term invoked to decribe the amount of surface area that has been modified by any given directional surface engineering method, e.g., such as *power beam surface alloying*, *peening* or *ion implantation.*

surface engineering. A term originally invoked by Bell and co-workers, circa 1983, to provide focus to the multi-disciplinary activities concerned with the science and technology of engineering surfaces. Its main goal is to make possible the design and manufacture of materials with a combination of bulk and surface properties unobtainable in a single monolithic material. The skill in surface engineering is to manipulate appropriate surface technologies to achieve optimal surface property designs, for specific applications, in the most cost effective manner.

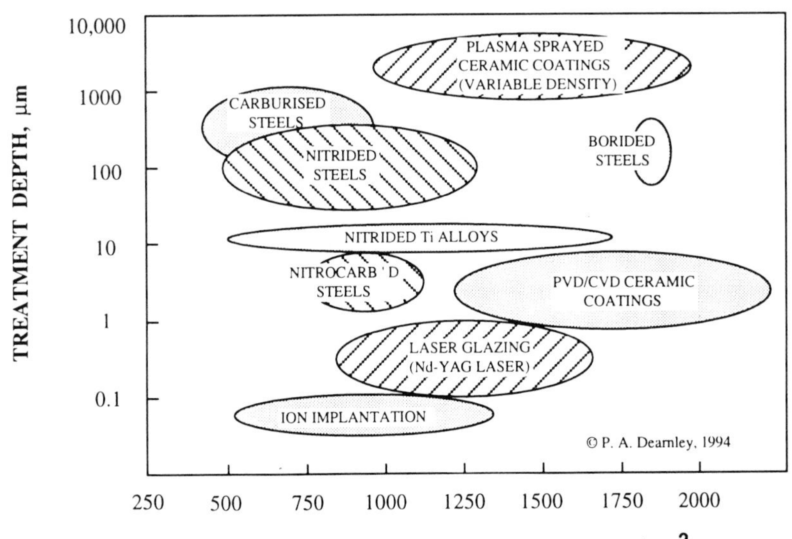

surface engineering charts. A convenient method of summarising the capability range of appropriate surface engineering treatments. For example, treatment depth can be plotted against hardness for treatments carried out on various substrates (see diagram). They may also be configured for a specific substrate or application area (e.g., see *titanium*).

surface finish. See *surface roughness.*

surface finishing. The chemical or mechanical processing step(s) which generate the final surface form of a component, e.g., finish turning or milling, diamond grinding and polishing.

surface hardening.

> 'Any treatment designed to render the surface of an object significantly harder' – IFHT DEFINITION.

Any treatment that causes an increase in surface hardness. Within the wider engineering community this term is frequently taken to mean induction hardening or carburising.

surface heating. See *surface heat treatment.*

surface heat treatment.

> 'Heat treatment aimed at changing the properties of the surface of the object under treatment' – IFHT DEFINITION.

A relatively old term, used in the heat treatment sector, to mean common thermochemical diffusion treatments like carburising, nitriding, nitrocarburising and boriding. The term does not usually refer to other methods of surface engineering, such as CVD, PVD or thermal spraying.

surface layer. A term usually meaning a coating, but sometimes referring to a surface residue resulting from corrosion or wear, e.g., metal cutting tools often become coated with surface layers of manganese sulphide after cutting free machining (high sulphur) steels.

surface modification. Any change in surface composition or geometry achieved by any means.

surface preparation. The procedure of pre-treating a surface prior to coating or diffusion, to remove surface oxides or contaminants. This may be achieved chemically, by acid pickling and/or fluxing, or mechanically by grit blasting. Also see *surface roughening.*

surface quench hardening. See *contour hardening* and *laser transformation hardening.*

surface rolling. See *roll hardening.*

surface roughening. A surface preparation designed to roughen the surface prior to thermal spraying or elctrodeposition, to effect a mechanical 'keying-in' or mechanical interlocking of the coating to the substrate. Usually achieved through grit blasting.

surface roughness. A parameter determining the numerical surface roughness. Measured, according to SI practice, as roughness average (R_a), although Centre-Line Average (CLA) and Arithmetic Average (AA) are terms still invoked by engineers in the English-speaking world. A surface roughness trace is obtained by a talysurf or similar instrument. For an idealised triangular surface roughness profile, R_a is 25% of the peak-to-valley height (R_t). For a surface generated by a single point turning tool:

$$R_a = 0.25f/(\tan A + \tan T)$$

Where f = the feed rate; A= tool approach angle; T= tool trail angle. In some circumstances the peak-to-valley height (R_t) is a more useful comparator of surface roughness.

surface tempering. In the context of carburising or carbonitriding, this refers to tempering the as-quenched martensitic case, typically conducted at 150–200°C, which serves to stress relieve the case while core properties remain unchanged.

surface working. See *roll hardening*.

surfacing. Another term for *weld hard facing* and any of the *thermal spraying* methods.

surfacing by welding. See *weld hard facing* .

surge plating. A method of electroplating in which the current is modulated periodically by superimposing surges or ripples or by adding an alternating current to the operating direct current

T

Tafel equation. Applicable to an electrode (anodic or cathodic) reaction where the charge transfer is rate determining. The equation predicts a logarithmic relationship between the activation overpotential (η_A) and current density (i):

$$\eta_A = a + b \log i$$

where *a* and *b* are the Tafel constants which are dependent upon the nature of the electrode process and the electrolyte.

target poisoning. A phenomenon in reactive sputtering whereby the sputter yield from a metallic target plate is substantially reduced when excess reactive gas is introduced into the sputter chamber causing the formation of ceramic compounds on the target surface. This is most commonly experienced when reactive sputtering titanium or zirconium targets with nitrogen in order to respectively produce titanium nitride and zirconium nitride. This phenomenon can be obviated by closed loop partial pressure control, which assures reactive sputtering below the critical partial pressure that results in poisoning. Also see *closed loop partial pressure control.*

tarnish resistance. The ability of a metallic surface or coating to remain lustrous in a normal atmosphere, i.e. not to become discoloured by atmospheric oxidation or sulphidation.

TD process. See *Toyota diffusion (TD) process.*

Teflon coating. See *PTFE coating.*

tempering. In the examples of carburised or carbonitrided steel cases, tempering is a simple heat treatment conducted at 150-200°C to relieve some residual stress and impart sufficient toughness to obviate brittle fracture.

temper resistance. The ability of a carburised or carbonitrided steel case to resist hardness loss through tempering.

Teniferbehandlung. See *Tufftride Process.*

terne coating. Any Pb-Sn alloy coating, containing up to 25 wt-% Sn, produced by hot dipping.

texture. The phenomenon of preferred crystallographic orientation (texture) often displayed by polycrystalline diffusion layers or surface coatings. These are typically of fibre texture character. The double-FCC TiN coating deposited by plasma assisted PVD methods, frequently, but not always, has the majority of its crystals growing in the [111] direction, while TiN coatings produced by CVD, frequently, but not always, grow in the [220] direction.A Pole figure for a CVD coating is given below. The TiN coating was deposited onto a 'steel cutting grade' cemented carbide substrate, and depicts a classical fibre texture. The intensity of the diffracted X-rays are expressed in 'times random' units, where 1 times random equates to a randomly oriented sample. LR and TR respectively refer to the longitudinal and transverse reference positions on the sample surface. Many other surface treatments can produce similar effects, e.g., the orthorhombic η-Fe_2Al_5 phase, formed during hot dip aluminising of steel, predominantly grows in the [002] direction. The practical implications of crystallographic texture, for example, in regard to wear resistance, are of clear importance, but have yet to be convincingly investigated. Texture is also sometimes referred to as preferred orientation. Also see *pole figure.*

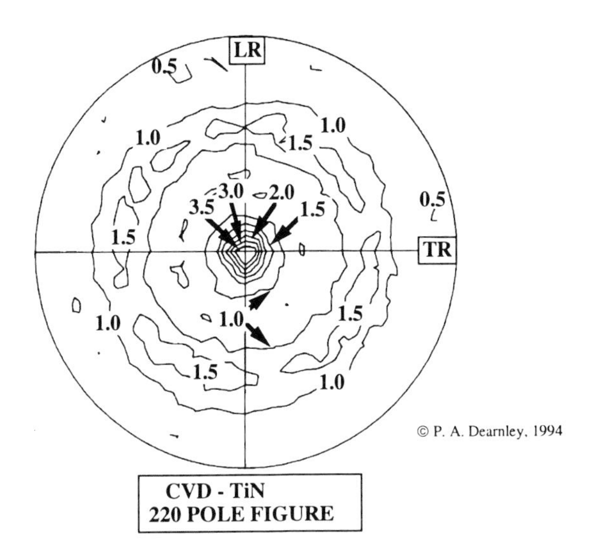

thermal barrier coating. A coating deposited onto components or structural elements exposed to high in-service temperature (e.g., combustion chambers, gas turbine blades and exhaust nozzles) with the aim of thermally insulating the substrate and protecting it from oxidation and/or sulphidation. One such coating comprises two layers, namely of a creep resistant bond coat of nickel-cobalt-chromium-aluminium-yttrium alloy and an outer coating of stabilised ZrO_2 or $MgO.ZrO_2$ (magnesium zirconate). Also see *thermal shock resistance.*

thermal shock resistance. The ability of a material to withstand an instantaneous change in temperature without cracking. This is an important parameter for ceramic coatings applied to gas turbine engine blades by, for example, plasma spraying. For elastic solids like ceramic coatings, the thermal stress (σ_T) generated in the coating by an instantaneous temperature change ΔT is given by:-

$$\sigma_T = E\alpha\Delta T/(1-\nu).\ B \qquad (1)$$

where E = Young's modulus (GPa); ν = poissons ratio (dimensionless); α = thermal expansion coefficient (°C^{-1}); B = Biot modulus (dimensionless).

Beyond a critical temperature change (ΔT_c) the thermal stress reaches the fracture strength (σ_f) of the ceramic coating. Hence, ΔT_c can be found by rearranging equation 1 and by replacing σ_T by σ_f:

$$\Delta T_c = \sigma_f(1-\nu)/E\alpha.B \qquad (2)$$

From this expression it can be seen that low modulus coatings are more tolerant to larger thermal shocks (i.e., ΔT_c is larger). This effect is exploited by developing microporous coatings, e.g., through plasma spraying which reduces the magnitude of E. Also, ZrO_2 has a relatively low intrinsic modulus; it is therefore a very useful coating for thermal shock or thermal barrier applications. It should be appreciated that rapid cooling is worse for ceramic coatings than rapid heating. In the former case the thermal stress is tensile in nature, enabling easy fracture propogation, in the latter, it is compressive and fracture propogation is more difficult. Also, the expansion characteristics of the substrate have to be taken into account when designing shock resistant thermal barrier coatings, and appropriate bond coatings should be selected.

thermal spraying. A general term covering many processes whereby a solid rod or powder of metal and/or ceramic is melted and propelled at an object or workpiece and resolidified. The resulting coating or spray deposit is built-up as an array of overlapping splats of rapidly solidified metal and/or ceramic. Thermal spray processes include; *thermospraying, plasma spraying (APS and VPS), detonation gun spraying, HVAF, HVOF, arc spraying* and *wire spraying.*

thermal stability. Immunity of a surface treatment to degradation by heat. For example, carburised cases loose strength when receiving prolonged exposure to temperatures above 250°C, while nitrided cases only start to degrade at temperatures exceeding 350°C. However, there are many specialised coatings that have been designed to provide protection above 750°C, e.g., those used in gas turbines. See *thermal barrier coatings.*

thermal stress. See *thermal shock resistance.*

thermionically assisted triode PAPVD. See *triode ion plating, triode sputter deposition* and *supported glow discharge plasma.*

thermochemical plating. See *chemical vapour deposition.*

thermochemical treatment.

> 'Heat treatment carried out in a medium suitably chosen to produce a change in the chemical composition of the object by exchange with the medium' – IFHT DEFINITION.

Any method of surface modification requiring heat and chemical action. All diffusion methods fall into this catagory. These can involve interstitial diffusion, as in carburising, nitriding and boriding, or substitutional diffusion as in chromising and aluminising.The term is also applicable to CVD.

thermoplastic polymer coatings. Polymer coatings which maybe softened in repeated stages, as many times as required, by appropriate reheating. The following thermoplastic coating materials are in common use: polyvinyl-organosol, polyvinyl-plastisol, polyvinylfluorides, vinyls and acryls (in water-dilutable and laminate form). Generally, thermoplastic coatings are more flexible but less abrasion resistant than thermosetting polymer coatings.Contrast with *themosetting polymer coatings.*

thermo-reactive diffusion process. See *Toyoto diffusion process.*

thermosetting polymer coatings. Polymer coatings that once solidified cannot be remelted. The following thermosetting polymer coating materials are in common use: alkyds, epoxies, acryls (for dispersion coatings), and polyesters. Generally, thermosettings are glossy, hard, brittle and chemically resistant.

thermospraying. One of the simplest forms of spray deposition, whereby powdered coating material is gravity fed or aspirated into an oxyacetylene flame. For metal or alloy powders, particle melting is very inefficient and the resulting coatings can be very porous. A number of special coatings have therefore been developed which undergo an exothermic reaction during their passage through the flame, typical of these is the 82:18 nickel-aluminium powder. This is a nickel powder that is coated with aluminium; it is not an alloy. The aluminium initiates an exothermic (thermite-type) reaction during spraying resulting in a coating of higher density than otherwise possible. Thermospraying is also used as the first step in laying down *fusion hard facing alloys* by the spray and fuse process.

Thornton diagram. A semi-schematic graphical depiction showing the influence of chamber pressure and homologous coating temperature (during deposition) on coating morphology. The diagram is applicable to metallic or ceramic coatings produced by plasma assisted PVD and encompasses some of the morphologies originally reported by B. A. Movchan and A. V. Demchishin (*Phys. Metal. Metallog.*, 1969, **28**, (4), 83-90.) for Vacuum deposisted coatings produced at different homologous deposition temperature. John Thornton discovered the existance of an additional morphological region 'Zone T' (for sputter deposited coatings) and characterised the influence of deposition pressure on the stability of the morphological zones. In general, it is preferable to produce most PVD coatings of the Zone T (fine-columnar) or Zone 2 (coarse-columnar) types; both are near fully dense. PVD-ceramic coatings are invariably of these types when substrate temperatures are insufficient to produce the equi-axed Zone 3 type morphology. Zone 1 coatings have a coarse-columnar morphology and contain significant levels of intergranular porosity and are to be avoided for most engineering purposes. The diagram shown is adapted from J. A. Thornton, *Ann. Rev. Mater. Sci*., 1977, **7**, 239-260.

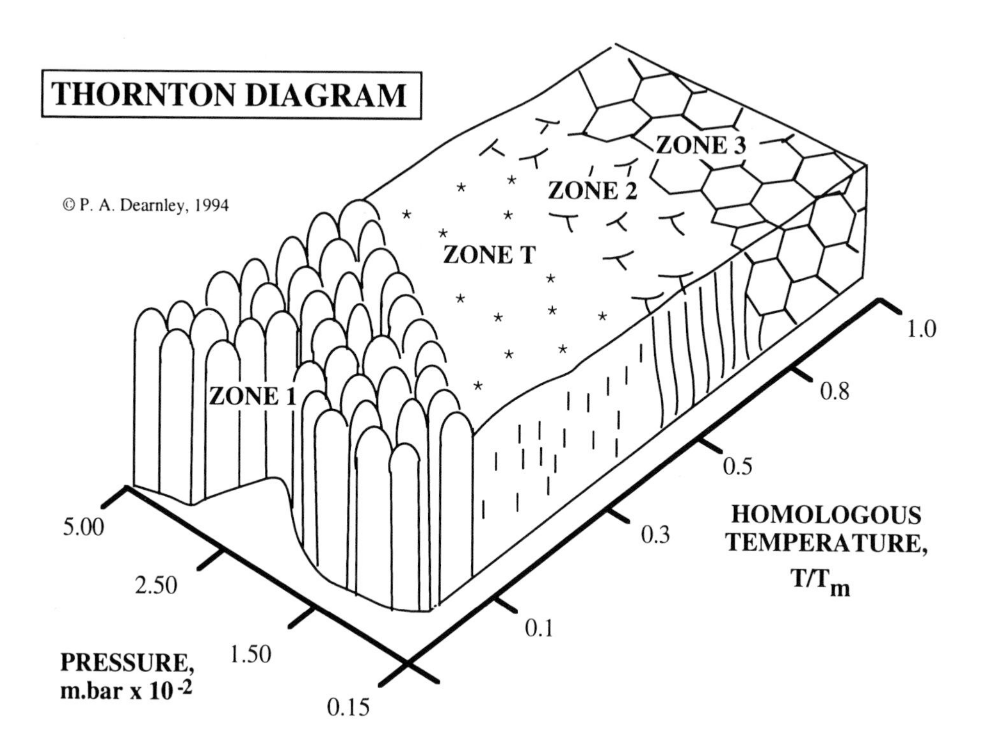

three-body wear. A situation arising when (usually) hard particles are trapped between two counterfacing sliding surfaces moving in different directions. A fluid (gas or liquid) may also be present. The particles cause abrasive wear of the sliding surfaces.

threshold limit value (TLV). Refers to the maximum time-weighted concentration for a 7 to 8 hour working day, assuming a 40 hour working week, that an individual can be exposed to any given substance. To be used strictly as a guideline to health hazards and not as a demarkation line between safe and dangerous exposure levels.

throwing power. The ability of a process to deposit a coating onto substrate surfaces placed in non-line-of-sight locations.

TiC. See *titanium carbide coating.*

TiN. See *titanium nitride coating.*

tin-copper plating. See *bronze plating* and *speculum plating.*

tin-lead plating. Two principal alloys are deposited. A 93Pb:7Sn (by weight) alloy used for bearing applications and a 60Sn:40Pb (by weight) alloy used for electrical components that require subsequent soldering. The coatings are electrodeposited from tin-lead fluorbate solutions contained in PVC lined tanks; the anode material is of the same alloy content as the required electrodeposit and a cathode current density of 3.2 amp/dm^2 at 1-3 volts is usually maintained during electrodeposition. The corrosion resistance of lead-tin plating is usually no better than unalloyed tin, but this alloy can offer some cost savings in some chemical plant applications such as wash boilers. Also see *bearing shells.*

tin-nickel plating. These coatings are electrodeposited from an electrolyte containing stannous chloride, nickel chloride ammonium bifluoride and ammonia. The most useful deposit has a composition of 65Sn:35Ni (by weight) which produces the NiSn intermetallic compound with a Vickers hardness of approximately 700 kg/mm^2. The residual stress state in the deposit is tensile when electroplated from new electrolyte; after some use, the residual stress state in subsequent deposits is compresssive. The NiSn deposit is more corrosion resistant (noble) than pure nickel and does not tarnish on prolonged atmospheric exposure, retaining its rose-pink lustre, even if the atmosphere contains relatively large quantities of sulphur dioxide and hydrogen sulphide. Hot caustic solutions (>10%) attack the coating only slowly. This coating is under exploited being presently reserved for domestic fittings and scientific instruments.

tinning. A hot dip method of coating iron or steel with tin. See *single-pot tinning*, *two-pot tinning* and *wipe tinning.*

tin plating. There are two principal processes: (i) electroless tin plating and; (ii) electrolytic tin plating. The former is applied exclusively to brass-plated, brass or copper objects. One method of electroless tin plating is known as the "boiling white process", which best serves small articles; they are mixed with granulated tin in an enamel lined vessel called a boiler. The contents are then covered with a 25g/l solution of potassium hydrogen tartrate and heated to boiling after which the bath is allowed to gently simmer. In this way the objects become tinned. After plating is completed, the objects are thoroughly washed and rinsed and, if necessary, polished by tumbling. An alternative electroless solution, comprises 50 g/l sodium cyanide, 5 g/l stannous chloride and 5.6 g/l sodium hydroxide. This procedure, carried out at ambent temperature, is used for coating circuit boards or relatively large objects.

In the electrolytic method, tin is deposited from either acidic or alkaline solutions. In the acidic bath process, a stannous sulphate-suphuric acid based elelctrolyte is used in conjunction with pure tin anodes which operate below 2 amps/dm^2, while the cathode current density is usually held at ~1amp/dm^2 at 0.6 to 0.9 volts. Tin plated steel is still used for the corrosion protection of steel cans used as food containers, despite recent success with alternative materials such as aluminium alloys.

tinting. Sometimes called heat tinting. A method of developing a range of aesthetic surface colours through flash oxidation. The technique is commonly applied to art work or jewellry made from titanium.

tin-zinc plating. In principle many tin-zinc alloys can be electrodeposited, however, only those containing 20-25 wt-% zinc have commercial importance. The coating has an attractive white finish and is quite ductile. It is deposited from an electrolyte comprising sodium stannate, zinc cyanide and sodium cyanide with a tin and zinc content of 30 g/l and 2.5 g/l respectively. The electrolyte is contained in a PVC lined tank held at ~65°C. Tin-zinc alloy anodes containing 73-77 wt-% Sn are favoured. The cathode current density is in the range of 1.1 to 3.2 amp/dm^2 at 3.5 to 4.5 volts. These coatings have found a niche application on steel that is to be used in contact with aluminium alloys, it serves to inhibit bimetallic corrosion; it neither stimulates corrosion of aluminium nor does it become consumed itself. In this respect it is superior to a galvanised surface. The electrodeposit's protective qualities can be increased still further by passivating for 10 to 15 seconds in a 2% chromic acid solution. This is followed by thorough rinsing in water.

titanising.

'Diffusion metallising with titanium' – IFHT DEFINITION.

The aim of this treatment is to increase the wear resistance, surface hardness, corrosion and cavitation resistance of steels and non-ferrous alloys. See *plasma assisted CVD* and *minor thermochemical diffusion techniques.*

titanium and its alloys. One of the more challenging groups of materials to surface engineer. Ion implantation has been advocated for medical implant devices, such as for the bearing femoral head surface of total hip replacements. After many years of *in-vivo* use ion implantation has proven unsatisfactory. Presently, alternative bulk materials are being used, such as solid aluminium oxide ceramics, although the potential of TiN coatings, deposited by PVD methods, is presently receiving investigation. The use of diffusion treatments, like nitriding, result in only a shallow hardened region (~5-10µm thick) after at least 24 hours at temperatures ≥750°C. These have proven of limited benefit. Recently, a hybrid treatment has been developed whereby PVD is carried out at ~700-750°C; this incorporates a nitriding treatment sequence prior to deposition of the usual TiN coating. The method has given highly satisfactory results. Much deeper surface treatments (≥1 mm) can be conveyed by laser alloying and interesting results have been obtained by aloying with nitrogen, boron or silicon. A difficulty with many of these treatments is that fatigue-life is often impaired. The main surface engineering methods for these materials are summarised in the diagram.

SURFACE ENGINEERING CHART; TITANIUM ALLOYS

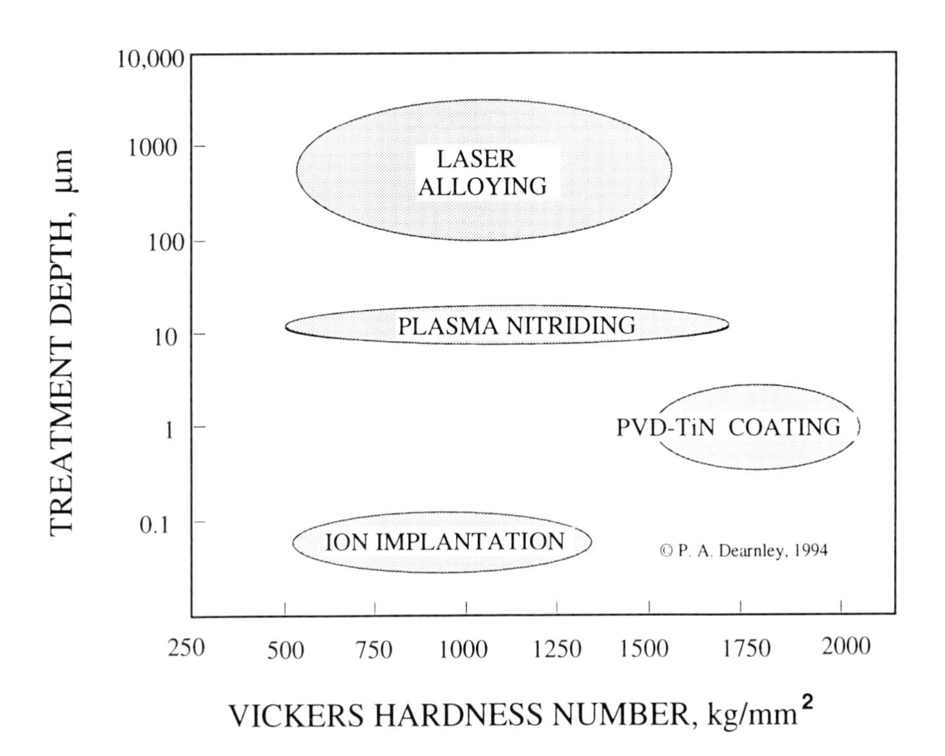

titanium carbide coating. One of the most common thin (~5 µm) hard (~2,000-3,000 Vickers) coatings applied to metal cutting and forming tools to convey wear resistance; usually applied by chemical vapour deposition (CVD). Used most commonly as one constituent of a multi-layered, multi-phased coatings, as discussed in the comments for *titanium nitride coating.* Also see micrograph in ***colour section,*** p. A . In respect to the citation for chemical vapour deposition (CVD).

titanium nitride coating. One of the most common thin (~5 µm) hard (~1800-2,500 Vickers) coatings applied to metal cutting and forming tools to convey wear resistance. Depending upon the substrate material it can be deposited by Chemical Vapour Deposition (CVD) or one of the plasma assisted physical vapour deposition (PVD) methods. It has moderate oxidation resistance and early investigations suggest it has a bio-compatibility similar to titanium, making it a likely future protective coating for the bearing surfaces of artificial hip and knee joints. The coating was first used in the 1970s for the wear protection of cemented carbide metal cutting tools, where it is still widely exploited, usually as a major constituent of a multi-layered, multi-phased coating. One variant comprises overlapping layers of TiC, Ti(C,N) and TiN. Such complex coatings are most easily applied by CVD. See micrograph in ***colour section***, p. A.

tool wear. Usually referring to the wear of metal cutting tools. See *diffusion wear, attrition wear, notch wear* and *discrete plastic deformation.*

torch hardening. See *flame hardening.*

total diffusion depth.

> 'Distance from the surface of a thermochemically treated object to the case/core interface, as specified by a characteristic of the material, e.g. hardness' – IFHT DEFINITION. Also see *case depth.*

total diffusion layer.

> 'Complete outer region of an object within which the composition has been changed as a result of thermochemical treatment' – IFHT DEFINITION Also see *case depth*

toucan indentation hardness. See *Knoop hardness.*

toxicity. The capacity of some substances to exert a highly harmful (poisonous) influence on man. This property should be taken into account in connection with cadmium plated vessels.

Toyota diffusion (TD) process. Also termed the thermoreactive diffusion-deposition process or TRD process. A reactive salt bath immersion process performed under normal atmosphere. Parts (commonly tool steels) are immersed into a borax based salt bath containing 20 wt-%V_2O_5 or 10 wt-% ferro-vanadium plus 5 wt-% B_4C. Compounds containing other carbide-forming elements like chromium can be chosen, as an alternative to vanadium, although vanadium is the preferred element for the protection of tool surfaces. During salt bath immersion, a carbide forming element is deposited on the surfaces of the parts and subsequently undergoes a diffusion reaction with carbon from the steel and forms a carbide layer. Hence, for example, vanadium reacts to form a monolithic layer of VC/V_4C_3. The technology was first developed by the Japanese Company Toyota and has had some implementation in Western Europe and North America. When set up to produce vanadium carbide coatings it can be regarded as a form of salt bath vanadising. Parallel techniques have been developed which utilise fluidised bed technology.

transferred arc. A special mode of operation of a plasma spray torch, in which one pole of the torch is switched to the object requiring treatment. The bias on the substrate can be either positive or negative; known respectively as positive transferred arc and negative transferred arc. It is usual to initiate the transferred arc mode at the commencement of a treatment to pre-heat the object(s) and drive-off any residual volatile contaminants. Temperatures above 300°C can easily be achieved. See ***colour section***, pp. F and G.

transformation hardening.

> 'Heat treatment comprising austenisation followed by cooling under conditions such that the austenite transforms more or less completely into martensite and possibly into bainite'
> – IFHT DEFINITION.

transmission coefficient, T. The ratio of visible light (or other electromagnetic radiation), transmitted through a given body, to that incident upon it.

transparency. The ability of a substance to transmit light, incident normally to the surface, without changing its direction. Some materials, e.g. aluminium vacuum coated glass, of sub-micron thickness, are capable of transmitting light.

TRD process. See *Toyoto diffusion process.*

tribological properties. A general term encompassing the frictional, lubrication or wear properties of a material.

Tribology. The science and technology of friction, lubrication and wear. A term first coined in 1966 by a British Government committee. From the Greek "tribos" (τριβοσ) which means rubbing.

triode ion plating. Any ion plating (evaporation source PVD) system that employs an auxiliary electron emitter such as a heated tungsten filament for raising the ion population in the glow discharge plasma that surrounds the objects being coated. This is usually augmented by a positively biased electrode. Overall, this approach increases the substrate ion current density, leading to an improved coating adherence. See *supported glow discharge plasma* and *ion plating.*

triode sputter deposition. Sputter deposition in which the target sputter yield is increased by raising the ion population through additional electron-neutral collisions. In this case the additional electrons are provided by a heated tungsten filamant. This can be augmented by a positively biased electrode. See *supported glow discharge plasma.*

trivalent chromium plating. A chromium plating process in which chromium is deposited from a trivalent chromium electrolyte. An inherently more efficient electrodeposition than from the usual hexavalent solutions, but technical difficulties prevented its use until 1975. Such baths are not yet widely used; further, they cannot be used for hard chromium plating.

Tufftride Process. A method of salt bath nitriding developed by the German company, Degussa, the novel feature of which is the deployment of a titanium crucible that obviates oxidation of the charge caused by solution of previously used steel crucibles and a subsequent decomposition of the sodium cyanide/cyante salts. Better control of the sodium cyanate content is also achieved by injecting air into the bath; this promotes the conversion of sodium cyanide to sodium cyanate. In Germany the process is termed 'Teniferbehandlung'. Also see salt bath *nitriding/nitrocarburising.*

tumbling. Also termed barrelling or rattling. A technique of deburring or descaling small components of regular shape by placing them in a rotating drum containing fine abrasive powder. Tumbling can be performed dry or wet. This technique is sometimes applied as a surface finishing operation and can be used to achieve honed edges of precise radii. For example, cemented carbide cutting tools are edge radiused in this way, prior to coating with ceramic layers (via CVD). Note the radiused cutting edge shown in the micrograph in the citation for *eta-phase (η-phase) zone.*

tungstenising.

'Diffusion metallising with tungsten' – IFHT DEFINITION.

The aim of the process is to increase the resistance of steels to corrosion by concentrated hydrochloric and sulphuric acids. See *minor thermochemical diffusion techniques.*

two-pot tinning. In this method of hot dip tinning the objects are immersed in a salt bath covered in molten fluxing salts (as for one version of single pot tinnning) and then immersed in a second bath of molten tin covered in oil or grease. The second bath is at slightly lower tempeature (~235-270°C) than the first (~280-325°C). The final coating is of uniform thickness (up to 20 μm) with a surface finish superior to single-pot tinned products. Graphitic cast-irons should be nickel or iron plated prior to tinning to obviate poor adherence of the tin plate. Also see *single-pot tinning* and *wipe tinning.*

U

ultramicrohardness. See *nanoindentation hardness.*

ultrasonic degreasing. A process of removing organic compounds, such as grease or oil, from the surface(s) of an object by means of ultrasonic agitation in an organic solvent. Ultrasonic degreasing can be used in conjunction with immersion solvent cleaning or vapour degreasing which enhances the degreasing effect, especially in the case of intricate objects.

ultraviolet reflectance. The ability of a surface to reflect ultraviolet radiation. Also see *reflectance.*

unbalanced magnetron. The unbalanced magnetron is a magnetron sputter cathode in which the outer magnetic field is made slightly stronger than the inner field, i.e., the magnetic fields are unbalanced (diagram). This enables more ions to escape the cathode and so contribute to raising the current density at the negatively or RF biased substrate; this increases substrate temperatures, obvating the need for auxiliary heaters and improves coating adherence. The technique was devised by Dr Brian Window of CSIRO, Lindfield Laboratories in Australia during the mid 1980's. It is the most important devlopment in sputter deposition technology in recent years. See ***colour section*** p. H. Also refer to B. Window and N. Savvides, *J. Vac. Sci. Technol.*, 1986, **A.4,** 453-456 and 504-508.

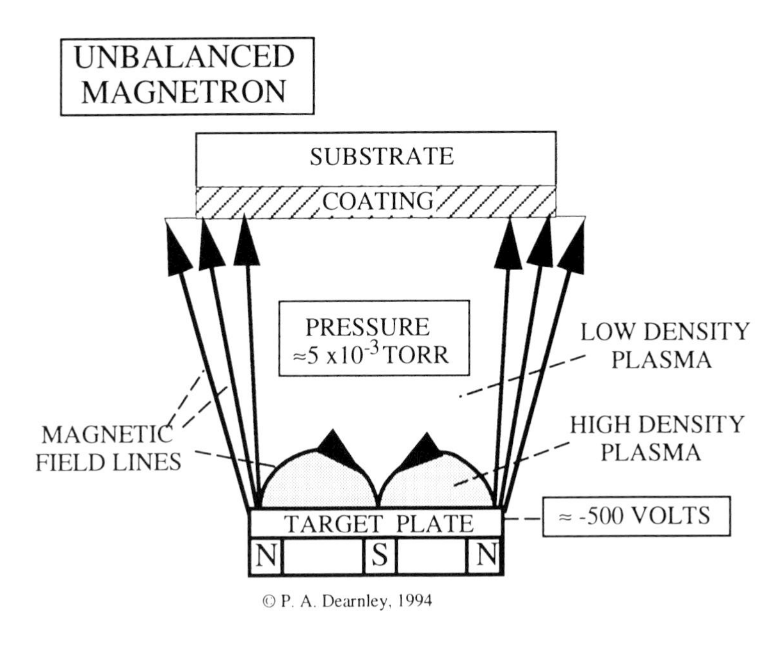

unbalanced magnetron sputter deposition. Sputter deposition using one or more unbalanced magnetrons. Also see *closed-field unbalanced magnetron sputter deposition.*

underlaying. The act of depositing a tough metallic layer onto the surface of an object prior to cladding by welding; minimises the risk of subsequent cracking and spalling of the clad layer.

underplate. An uncommon term referring to the electroplated coating(s) applied before the final (exterior) electroplating.

V

vacuum carbonitriding.

'Gaseous carbonitriding carried out at subatmospheric pressure" - IFHT DEFINITION.

vacuum carburising.

'Gas carburising carried out at subatmospheric pressure' – IFHT DEFINITION.

A system of carburising using the boost-diffuse approach (explained previously in connection with *plasma carburising*). Under a rough vacuum (0.1 to 0.3 torr) the steel charge is austentised in the range of 850 to 1050 °C, although there are economic advantages in working at the upper end of this range (shorter treatment times). The carburising (boost) cycle is carried out under an appropriate hydrocarbon gas (e.g., CH_4 or C_3H_8), using a partial pressure ~10 to 50 torr for graphite lined furnaces, or ~100 to 200 torr for ceramic lined furnaces. The susbsequent "diffuse" cycle is carried out at 0.5 to 1.0 torr after which the hot components are oil or gas quenched. The boost-diffuse cycle represents a significant deviation from equilibrium conditions.

Vacuum carburising vessels comprise two adjacent chambers, a vacuum furnace and a quench chamber, connected by an inner vacuum door. Furnaces should be capable of maintaining uniformity of temperature within ± 7°C. A system of pulse-pumping is used during the boost cycle to enable better uniformity of gas distribution. Gas is admitted until the desired partial pressure of hydrocarbon is obtained. This is then removed by the vacuum pumps and when a lower set point has been reached, further process gas is admitted and the cycle is repeated. For boost-diffuse cycles conducted above 1000°C, it is common practice to allow the charge to cool to 900°C before quenching, to minimise dimensional distortion.

vacuum coating. Also called evaporation coating. A method of applying thin metallic or ceramic coatings to substrates at low temperature. The process is usually carried out under high vacuum conditions (~10^{-5} to 10^{-6} torr). The coating material can be vapourised using a variety of heat sources; the most common are resistance heaters for low melting point materials, like aluminium, zinc sulphide and antimony, and electron beam heaters for high melting point materials like nickel, molybdenum and aluminium oxide. The process is highly directional (diagram) and the quality of coating adherence is poor compared to those produced by ion plating. Hence, such coatings have limited tribological application. Some coatings like aluminium oxide and zirconium nitride can be produced reactively by evaporating the metal constituent in a partial pressure of oxygen and nitrogen respectively. For reactive deposition the chamber pressure maybe as high as 10^{-2} torr. Designs also exist that permit two or more metals to be evaporated simultaneously, enabling the synthesis of alloy coatings. Aluminium coatings are widely used for decorative purposes on many substrates, including plastics. They are also widely used in the auto industry for mirrors and lamp reflectors. In the electronics industry capacitors are fabricated using coatings of the oxides of cerium, silicon, tantalum, titanium and aluminium, while resistors comprising composite coatings of chromium plus silicon dioxide are frequently used. The latter undergo a post-deposition heat treatment at ~430°C to develop the required resistance. The majority of vacuum coatings are less than 1 μm thick.

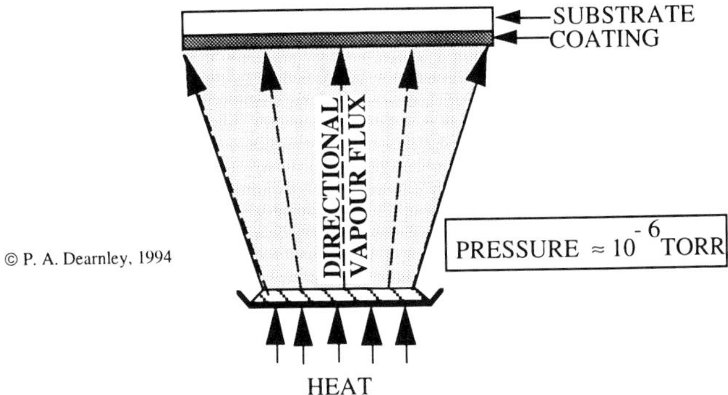

vacuum nitriding.

'Gas nitriding carried out at subatmospheric pressure' – IFHT DEFINITION.

vacuum plasma spraying (VPS). Also called low pressure plasma spraying (LPPS). A plasma spraying process performed in a chamber that is firstly evacuated to ~10^{-2} Torr and secondly progressively back filled with an inert gas (usually argon) to ~100 Torr (diagram). The plasma torch is often "warmed-up" before the latter working pressure is reached. VPS has gained popularity because the resulting coatings are denser (>90% of theoretical) than those achieved during conventional plasma spraying, conducted at atmospheric pressure. The latter method is sometimes called 'APS'; air plasma spraying. See ***colour section*** pp. F, G and H.

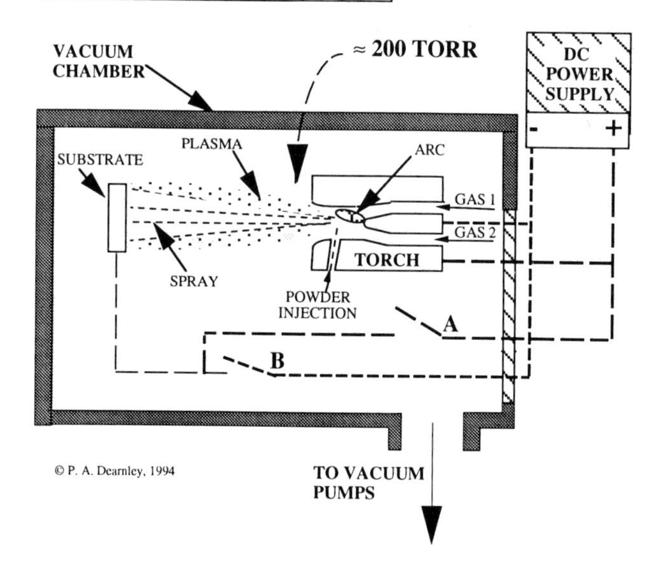

vanadising.

'Diffusion metallising with vanadium' – IFHT DEFINITION.

The aim of the process, applied to iron and steel, is to increase their wear and corrosion resistance, especially to 50% HNO_3, 85% CH_3COOH, 10% NaCl or 98% H_2SO_4. It is usally carried out using salt-bath or pack media above 900°C. One important variant of this method is the *Toyota diffusion (TD) process.* Also see *minor thermochemical diffusion techniques.*

vapour blasting. See *wet blast cleaning.*

vapour chromising. A rare form of chromising, whereby chromium is heated in-vacuo to 1050 to 1200°C and slowly sublimes. The vapour condenses onto the component surfaces and is subsequently dissolved. A more pragmatic approach is to use conventional chromising techniques. See *gaseous chromising* and *pack chromising.*

vapour degreasing. Degreasing of a surface achieved by exposure to the action of vapourised organic solvents. Maybe combined with spraying or immersion in hot liquid solvent in conjunction with ultrasonic agitation.

vapour deposition. See *vacuum coating, plasma assisted PVD, PVD* and *CVD*

vibratory finishing. A surface finishing operation in which the objects under treatment are embedded in a vibrating container containing an abrasive medium. Also see *tumbling.*

Vickers hardness. Also called diamond pyramid hardness (DPN). A square based pyramid indenter, made of diamond, whose apex has an included angle of 136° is used to indent the surface of a metal or ceramic under the application of a known load. Following indentation, the average indentation diagonal (*d*) is measured and the Vickers Hardness Number (VHN) obtained from tables or calculated. VHN is the ratio of the load (*P*) to surface (projected) area of the indentation, and is usually expressed in kg/mm^2. VHN can be calculated using the formula:

$$\text{VHN} = \text{load/contact area} = 1.854\ P/d^2 = \text{kg/mm}^2$$

Nowadays, it is becoming popular to express Vickers hardness in GPa; this is achieved by expressing *P* in Newtons rather than kilogrammes (1 kg = 9.80665 N).

VHN is widely used in the heat treatment and surface engineeering sectors and both macro and microindentation versions exist. Vickers microhardness is especially useful for obtaining hardness-depth profiles of surface engineered materials like nitrided and carburised steels. For example, see *hardness profile.*

vinyl coating. A plastic dispersion coat containing themoplastic copolymers of polyvinylchloride (PVC) and polyvinylacetate (PVA). It has excellent formability and is widely used as a protective finish for steel sheeting used in deep drawing applications.

vitreous enamel coating. A strongly adherent glassy substance applied to (mainly) steels and grey cast-irons as an inexpensive abrasion and corrosion resistant coating. Vitreous enamel is prepared by smelting together an intimate mixture of refractory materials. For example, silica, titania, feldspar and china clay are mixed together with an appropriate fluxing agent such as borax, sodium silicoborate and nitrates or carbonates of lithium, sodium and potassium. Melting together of these materials produces a substance resembling glass in texture, but often containing gas bubbles. The material is then rapidly solidified, e.g., by pouring into water, then crushed and ball milled into a fine powder called "frit". Cast-iron surfaces are prepared by grit blasting, while steels are pickled in acid and given a nickel dip or hot dip aluminised coating. Frit is sometimes applied to cast-iron surfaces by sieving onto the surface and heating the objects in a furnace to 900°C; this is called the 'dry process'. A more common approach is to apply the frit as a slurry by spraying or dipping; after drying, the frit is fused by heating to ~750-850°C. Items are then allowed to cool slowly. Formerly in wide-use for cook-ware and other house-hold implements, enamelling is nowadays largely superseded by PTFE coatings.

VPS. See *vacuum plasma spraying.*

W

washing. The process of removing surface residues by means of clean water or water containing passivating, wetting or detergent additives. It may be carried out in running or stagnant water, either hot or cold, by immersion or spraying.

Watts bath. The most commonly used nickel plating solution, composed of nickel sulphate (240-330 g/l), nickel chloride (37-52 g/l) and boric acid (30-40 g/l).

WDS. Wavelength dispersive spectrometry; the same as WDX.

WDX. Wavelength dispersive X-ray analysis. Sometimes available as an 'add-on' to a standard SEM. Formerly known as "electron microprobe analysis" (EPMA). An electron beam (~20-30 keV) is focused onto a conducting surface and among the many surface effects, X-rays, of characteristic wavelength are produced. The optimal sample-electron beam distance is obtained with the aid of a light-optical microscope; this ensures that the X-rays are brought to focus at the X-ray counter where they are collimated and diffracted by a crystal spectrometer. The geometry of the crystal is configured to enable diffraction according to the Bragg equation:

$$n\lambda = 2\,d\,\sin\theta$$

where λ is the X-ray wavelength, d is the interplanar spacing (of the crystal), n is the order of reflection and θ is the angle of diffraction (Bragg angle). Accordingly, within a given wavelength range, the crystal is 'tuned' to the incoming X-rays by varying θ; this is achieved by rotating the crystal. Several crystals are needed to cover the range of 6.76 nm for B K_{α} to 0.013 nm for U $K_{\alpha 1}$. In this way, most elements can be detected. The intensity of the diffracted X-rays is proportional to the amount of a given element in the sample, which can be quantified after applying a 'ZAF' correction; i.e., after accounting for the effects of sample atomic number (Z), X-ray absorbtion (A) and X-ray fluorescence (F). For quantification it is essential to determine the peak count rate and background count rate for an external primary standard, ideally for the pure element. Elemental mapping, in conjunction with secondary electron imaging is also a useful feature of this method. For a recent review see: M. G. Hall, *Surf. Eng.*, 1993, **9**, 205-212.

wear. The loss of material from a surface when brought into rubbing contact with two or more other surfaces; usually measured as weight loss, volume loss or dimensional change. The type and amount of wear is determined by many factors. The most important are atmosphere, contact stresses, contact temperatures, contact speed, direction of contact and presence of third-body particles. The nature of the wear interface is of crucial importance. Also see *seizure* and *lubrication*.

wear equations. See *Wear Theory* and *Archard's wear equation.*

wear life. Service life in terms of material loss through wear; sometimes expressed as the time required to achieve a critical dimensional change or weight loss.

wear properties. A general term denoting the response of any given material to wear. This should be qualified by specifying the wear environment and the rate controlling wear mechanisms.

wear resistance. The ability of a material to endure wear.

wear theory. There is no unified theory of wear. Theories exist to explain and predict specific types of wear, e.g., Quinn's oxidational wear theory or Suh's delamination theory. In the past such theories have been hampered by a dearth of light- and electron-microscopy and other surface analytical techniques. This has only recently begun to be addressed. However, knowledge of the temperture(s) and stress(es) acting at wearing surfaces remains poorly quantified. Many observations, if made at all, are often poorly executed and weakly interpreted. This contrasts with the large amount of available data concerning dimensional or weight changes caused through wear. In the absence of rigorous observation, such data has limited value, impairing theoretical development.

weather resistance. The ability of a material to withstand degradation through the action of the natural atmosphere, i.e., rain, snow, frost, wind, sunlight etc. This action manifests itself as changes in reflectance, colour, and freeness from surface pores, especially in regard to polymeric coatings.

weld hardfacing. A technique for laying down very thick (~1 to 10 mm) layers of wear resistant material. Various welding techniques can be used, these include metal-inert gas (MIG), tungsten-inert gas (TIG), plasma transferred arc (PTA), submerged arc and manual metal arc. The latter, because of its relative simplicity, is found in widest use. A very broad range of coating materials can be applied. They include stellites (Co alloys), martensitic and high speed steels, nickel alloys and WC-Co cemented carbides. After deposition, it is frequently necessary to finish machine the object to size by metal cutting or grinding. Also see *fusion hard facing alloys.*

wet blast cleaning. Also called vapour blasting. This is an abrasive blast cleaning process in which fine non-metallic particles are propelled against the surface to be cleaned by a liquid stream (usually water). This process is less aggressive than dry blast cleaning; it is primarily reserved for fine finishing or for preparation of surfaces prior to coating.

wet galvanising. Hot dip galvanising in which the objects to be treated are not prefluxed after pickling but are immersed in molten zinc by passing them through a floating flux blanket. Contrast with *hot dipgalvanising.*

wettability or wetting. The extent to which a solid surface is wetted by a liquid; usually expressed as wetting angle. A zero degree wetting angle corresponds to complete wetting, while a high angle of 70 or 80° corresponds to poor wetting.

white layer. See *compound layer.*

wipe test. A technique for determining the optimal plasma spray processing parameters that produce the maximum particle melting efficiency during plasma spraying. It was originally devised as a simple method for estimating the melting efficiency of various plasma source gases. A modified version of the test is schematically depicted. The plasma torch is rapidly traversed across, and parallel to, a plate containing a 5-10 mm wide vertical aperture (placed approximately 150mm in front of the torch). The distribution of the spray particles that pass through the aperture are collected on a metal foil attached to a rotating steel cylinder (the direction of rotation being opposed to that of the traverse).

The test expands the horizontal plane of the spray flux, allowing a relatively easy assessment of melting efficiency and other phenomena to be made, which at the same time takes account of particle trajectory history; the individual melted particles, or "splats" are examined by scanning electron microscopy. (Note that it is particularly important to arrange for the injection port of the torch to lie in the same plane as that which is spatially separated by the test). The test easily discriminates between particles that have passed across (**A**) or along (**B**) the "principal axis" of the plasma (i.e., those particles which passed along or across the regions of highest temperature) from those which remained in the outermost and relatively cool portions of the plasma torch (**C**). Further details are given in P. A. Dearnley and K. A. Roberts, *Powder Metallurgy*, 1991, **34,** (1), 23-32.

WIPE TEST

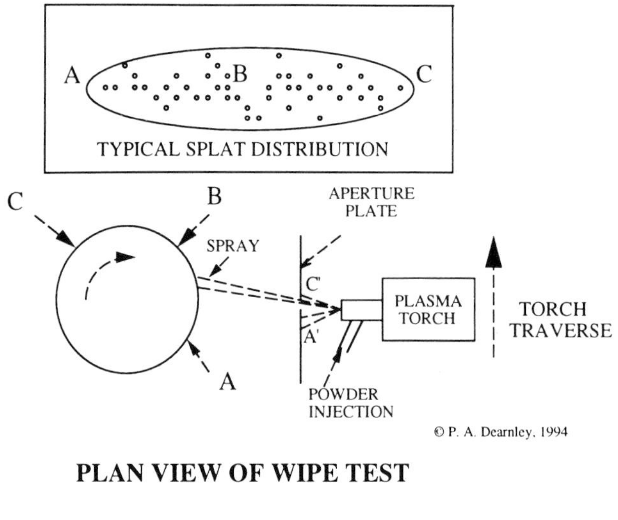

A B C

TYPICAL SPLAT DISTRIBUTION

C B APERTURE PLATE

SPRAY

C'

A'

PLASMA TORCH

TORCH TRAVERSE

A

POWDER INJECTION

PLAN VIEW OF WIPE TEST

wipe tinning. A method of tinning very large ferrous objects when the hot-dipping methods of single- or two-pot tinning are unfeasible, or if only selected areas of an object require coverage with tin. The process is relatively crude and involves removing passive surface oxides from the surface of the object by immersion in an aqueous flux and heating it with an appropriate gas burner to ~280-300°C; a small slug of tin may be placed in the immediate area which will commence melting when the correct temperature has been reached. Once melting commences, further tin is applied to the area and is spread over the surface with a steel scraper or wire brush. A soft fabric pad is then wiped over the surface to produce the final finish. It is important to thoroughly wash the treated area to remove excess fluxing agent, otherwise corrosion will result. Wipe coatings are up to 15 μm thick.

wire brushing. A manual technique for descaling metallic objects. The surface is subjected to the rubbing action of a hand held or motor mounted steel wire brush. The method is only suitable for the removal of loosely adhering scale.

wire coating. The process of coating copper wire with an insulating plastic layer. The wire is passed through a special wire coating die at speeds ranging from 1 to 1800 m/min. The die feeds molten plastic from a reservoir into a tunnel through which the wire passes. The plastic becomes 'wiped' onto the metal surface; the coated wire is then passed through a water filled cooling trough, which can be several hundred meters long, and then wound onto drums. The preferred polymers for primary electrical insulation are low or medium density polyethylene; nylon is used for jacketing because of its relative strength, toughness and abrasion resistance.

wire flame spraying. See *combustion wire gun spraying*

wire spraying. See *combustion wire gun spraying*

working life. See *service life.*

working properties. The ability of a surface engineered material to endure in-service wear, corrosion or fatigue. The same meaning as durability.

X

XPS. X-ray photoelectron spectroscopy. The surface under interrogation is irradiated by monochromatic low energy ('soft') X-rays, e.g., Al K_{α}, under ultra high vacuum conditions (diagram). This produces an effect termed photoionisation which leads to the emission of photoelectrons from the atomic cores of the surface atoms. The latter have a kinetic energy (E_k) that is related to the incident X-ray energy ($h\nu$) and the core binding energy of the electron (E_b) by the Einstein relation:

$$E_k = h\nu - E_b$$

In XPS, a kinetic energy distribution is obtained, such that a plot of intensity versus binding energy can be derived. Binding energies corresponding to, for example, 3d, 3p and 3s transitions are obtained. Their energies are unique to a given element, enabling precise elemental identification. It is also possible to use this information to gain chemical bonding information, e.g., to demonstrate whether a given element is bonded to another, or not. XPS can be used on non-conducting surfaces but cannot be generally applied to sample surface areas of less than 1 cm^2. However, at the time of writing, attempts are being made to improve the capability of XPS to enable selected area analysis.

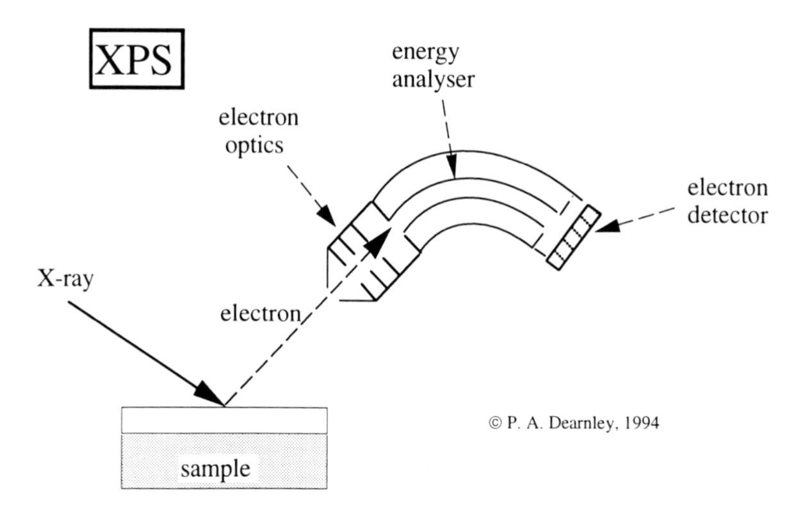

X-ray diffraction. A general purpose method of phase analysis and lattice parameter determination. The basis of X-ray diffraction is the Bragg equation:

$$n\lambda = 2d\sin\theta$$

where λ is the wavelength of the X-ray radiation, n is the order of reflection, θ is the Bragg angle and d is the inter-planar distance of the planes causing diffraction. By irradiating a sample surface with monochromatic X-rays (e.g., Cu $K_{\alpha 1}$) and measuring θ, d is readily determined. The intensity of the diffracted X-rays are also recorded. Nowadays, d-spacing and X-ray intensity data is obtained using an X-ray goniometer which uses a proportional counter to measure diffraction intensity; an output of intensity versus 2θ is obtained. For a randomly oriented polycrystalline solid, several diffraction peaks of characteristic intensity will be produced. The international JCPDS data base should be consulted when undertaking phase analysis; this is a major collection of diffraction data. Much greater detail on X-ray diffraction principles and methods are available in standard texts, e.g., see B. D. Cullity: *Elements of X-ray Diffraction*, 2nd ed; 1978, Addison-Wesley, Reading MA. Also see *glancing angle X-ray diffraction.*

Y

YAG laser. See *Nd–YAG laser.*

yttrium–aluminium–garnet laser. See *Nd–YAG laser.*

Z

Zinal. A proprietry surface treatment developed in France by Hydromécanique et Frottement (HEF) for the surface protection of aluminium alloys. It comprises a two-stage process involving: (i) electroplating an indium based coating onto the aluminium surface; (ii) an annealing treatment between 120 and 150°C during which copper and zinc are also introduced into the surface layer. The total depth of the surface layer is ~25 μm. Hard intermetallic compounds are formed in the layer, giving a Vickers hardness of around 400 kg/mm^2. The layer is claimed to improve resistance to galling or seizure and can improve lubricity in applications where organic lubricants cannot be used.

zinc plating. Also called electrogalvanising. An electroplating process in which zinc is deposited on the surface of an object from cyanide, alkaline non-cyanide, or acidic chloride baths. Pure zinc anodes are used. Zinc plating is applied mainly to steels for corrosion protection and decoration purposes, but the technique has poor levelling capability. Zinc plated coatings are pure, very ductile and up to 12 μm thick, compared with 80-125 μm for hot dip galvanised layers.

zirco-aluminising. The aim of the process is to increase heat (oxidation) resistance of iron and plain carbon steels. See *minor thermochemical diffusion techniques.*

zirconia coating. See *thermal barrier coating.*

zirconia diffusion coating. An external zirconia (ZrO_2) layer formed on the surface of zirconium as the result of controlled oxidation. Further details are given under *oxidation.*

zirconia oxygen sensor. Also called an oxygen probe. Used for detecting small levels of oxygen in gas carburising furnaces and thereby obtaining an estimate of carbon potential. From the following reaction:

$$CO \longrightarrow C_{(Fe)} + 1/2\ O_2$$

carbon potential can be seen to be proportional to $(pCO)/p(O_2)^{1/2}$. The zirconium sensor generates an e.m.f. when brought into contact with oxygen, using air as a reference. One design of sensor has the following relationship:

$$\text{e.m.f} = 4.9593 \times 10^{-5}\ T \log_{10} (pO_2/0.209)$$

where T is the absolute temperature of the sample point and pO_2 is the partial pressure of oxygen. Hence, the measured e.m.f may be directly related to carbon potential. Also see *gaseous carburising.* This method has recently received rigorous investigation for controlling gaseous nitriding atmospheres, see S. Böhmer, H. J. Spies, H. J. Berg and H. Zimdars, *J. Surf. Eng.*, 1994, **10**, (2), 129–135

Bibliography

Y. ARATA: *Plasma, Electron & Laser Beam Technology,* American Society for Metals , Ohio, 1986 .

T. BELL: *Survey of the Heat Treatment of Engineering Components,* The Metals Society, London, 1973 .

F. P. BOWDEN and D. TABOR: *The Friction and Lubrication of Solids,*Oxford, Clarendon Press 1986.

V. E. CARTER: *Metallic Coatings for Corrosion Control,* Butterworths, London, 1977.

R. CHATTERJEE-FISCHER (ed): *Nitrieren und Nitrocarburieren,* expert-Verlag, Sindelfingen, 1986.

H. C. Child: *Surface Hardening of Steel*, Oxford University Press, Oxford, 1980.

V. DEMBOVSKY: *Plasma Metallurgy,* Elsevier, Amsterdam, 1985.

J. K. DENNIS and T. E. SUCH: *Nickel and Chromium Plating,* 3rd edn, Woodhead Publishing, Cambridge, 1993.

L. C. FELDMAN and J. W. MAYER: *Fundamentals of Surface and Thin Film Analysis,* North Holland, New York, 1986.

P. D. HODGSON (ed): *Quenching and Carburising*, The Institute of Materials, London, 1993.

I. M. HUTCHINGS: *Tribology*, Edward Arnold, London, 1992.

K. L. JOHNSON: *Contact Mechanics*, Cambridge University Press, Cambridge, 1987.

E. MACHERAUCH and V. HAUK(eds): *Residual Stresses,* 1986, Oberursel, Deutsche Gesellschaft fur Metallkunde.

D. MATEJKA AND B. BENKO: *Plasma Spraying of Metallic and Ceramic Materials,* John Wiley & Sons, Chichester, 1989.

P. L. B. OXLEY: *Mechanics of Machining,* Ellis Horwood Limited, Chichester, 1989.

L. L. SHREIR (ed): Corrosion, 2nd edn, Butterworths, London,1976.

K-E. THELNING: *Steel and its Heat Treatment,* Butterworths, London, 1981.

E. M. TRENT: *Metal Cutting,* 3rd edn, Butterworths, London, 1991.

E. TYRKIEL (ed): *Multilingual Glossary of Heat Treatment Terminology,* The Institute of Metals, London, 1986.

J. L. VOSSEN and W. KERN (eds): *Thin Film Processes 2,* Academic Press, New York, 1991.

J.. M. WALLS (ed): *Methods of Surface Analysis,* Cambridge University Press, Cambridge, 1990.

R. B. WATERHOUSE: *Fretting Corrosion,* Pergamon Press, Oxford, 1975 .

The Canning Handbook, 23rd edn, W. Canning plc, Birmingham, 1982.

Wear Resistant Surfaces in Engineering, HMSO, London, 1986.